COSMIC MAN
The Divine Presence

COSMIC MAN
The Divine Presence

The Theology of St. Gregory of Nyssa
(ca 330 to 395 A.D.)

Paulos Mar Gregorios

A NEW ERA BOOK

PARAGON HOUSE
New York

First American edition, 1988

Published in the United States by

Paragon House Publishers
90 Fifth Avenue
New York, NY 10011

A New Ecumenical Research Association Book

Library of Congress Cataloging-in-Publication Data

Paulos Mar Gregorios, 1922-
Cosmic man.

Reprint. Originally published: New Delhi : Sophia Publications, 1980.
"A New Ecumenical Research Association book."
Bibliography: p.
Includes index.
1. Gregory, of Nyssa, ca. 335-ca. 394. 2. Man—History. 3. God—History of doctrines—Early church, ca. 30-600. 4. Cosmology—History. I. Gregory, of Nyssa, Saint, ca. 335-ca. 394. II. Title.
B667.G74P38 1988 128'.092'4 87-7199
ISBN 0-913757-91-8 (pbk.)

CONTENTS

PART TWO
GOD—CREATION—MAN

FOREWORD

The author of this beautiful book is a churchman, a scholar, and a monk. These roles are uniquely combined in a man of deep faith and intellectual daring. While this combination is unusual today, it is a human type that the Orthodox tradition has nurtured over the centuries. Paulos Mar Gregorios is the Metropolitan of Delhi and the North in the Syrian Orthodox Church of India, the Principal of the Orthodox Seminary in Kottayam, Kerala, a past Associate General Secretary of the World Council of Churches and one of its current Presidents. His distinguished career as a churchman and scholar is marked by an abiding passion for justice between nations, an openness to dialogue with people of different faiths and intellectual commitments, and a deep encounter with technological civilization. The culmination of these concerns is this book—a searching exploration of the very foundations of human existence. Allow me to trace its most significant elements.

This remarkable study by Paulos Mar Gregorios—monk, scholar, and churchman—brings an important new voice to contemporary theological discussion in North America. For too long the North American discussion has been dominated by Protestant, and more recently, Catholic voices. But that situation has begun to change in recent decades. And in this study we have a new voice, a Syrian Orthodox and Eastern voice, joining that small band of Orthodox voices, most notably G. Florovsky, A. Schmemann, and John Meyendorff, that has contributed so significantly to making the Christian conversation in North America more truly ecumenical. One aspect of that Orthodox contribution, discernible here too, is the conviction of the enduring relevance of the great thinkers of the tradition, in this case, Gregory of Nyssa, to the contemporary world. With Metropolitan Mar Gregorios' skillful handling of texts and lucid exposition of their content, Gregory of Nyssa becomes our contemporary.

The importance of the Metropolitan's contribution should not, however, be construed as limited to the narrow theological discussions among experts. His concern is broader and more audacious. He seeks, through his study of Gregory of Nyssa, to engage our technological civilization in an argument concerning underlying assumptions about humanity. Facing the technological civilization of our time, he writes, "It would be fatal for us . . . to go on oscillating between . . . an otherworldly mysticism . . . and a secular humanism that ignores the ground and source of the being of ourselves and the cosmos." Thus he challenges our technological civilization to reconsider the ontological questions we have ignored. He wishes to engage us all—scientists, humanists, modern men and women—in a fundamental inquiry concerning the most basic question of all: humanity's relationship to "the source and ground of its being." Here Gregory of Nyssa is his guide, and, according to Metropolitan Mar Gregorios, he could be ours as well.

The third aspect of this study worth noting here is its unapologetic philosophical character. As Metropolitan Gregorios states at the outset of his study, he aims to "analyze the problem of human existence" between the two poles of "God and creation." This aim is not, however, armchair philosophy nor the philosophical approach dominant in the academy today. It is rather, as the ancients knew, a quest for wisdom, for understanding ourselves with depth so that our lives might gain orientation and direction. This love of wisdom—the root meaning of philosophy—is everywhere apparent in these pages and reflects Metropolitan Gregorios' deep roots within his Orthodox and Indian traditions. As a monk his life has unfolded in the depths of a vital faith and within a still vital spiritual culture. These roots are evident in his study. Yet Metropolitan Mar Gregorios is not unfamiliar with Western philosophy and culture. Important aspects of his education were undertaken in North America, he has traveled widely with the World Council of Churches, and he acted as a principle force behind the important conference on "Faith, Science, and the Future" held at the Massachusetts Institute of Technology (M.I.T.) in 1979. Thus it is as an informed contemporary, deeply concerned about the path of our technological civilization, that he has turned to the fundamental questions of human life, in the hope that we might all find direction.

In *Cosmic Man, The Divine Presence*, then, you will find not an easy book but an important one. As Paulos Mar Gregorios leads us through the intricacies of Gregory of Nyssa's insights, he is seeking to lead us into the divine presence in our cosmic humanity.

M. DARROL BRYANT
Chair, Department of Religious Studies
Renison College, University of Waterloo

PREFACE TO THE AMERICAN EDITION

Why Gregory of Nyssa?

Why would a Christian (or anybody else for that matter) living in the twentieth century be interested in Gregory of Nyssa, who lived and wrote in the fourth century?

Let me try to advance a few reasons.

First of all, Gregory of Nyssa offers us an alternative way of understanding the Christian faith—a time-honored way, the respected way of the Eastern Orthodox tradition. It can do no harm, especially to westerners who find their tradition of Christianity rather problematic, to examine this alternative. Gregory does not go the way of Augustine, nor of Calvin, Luther or Thomas Aquinas. All these accepted western traditions could find a more creative restatement if they were exposed to Gregory of Nyssa.

Second, this most philosophical of Christian fathers addresses modern problems of ecology in his approach to life in this world as the Western fathers of the Church do not. Gregory's view of the relation between the human species and the rest of creation, and of both to God, can give us some new perspectives on this thorny problem of our advanced industrial civilization.

Third, Gregory has something very pertinent to say about human freedom, which must be of some significance to liberation theologians, as well as for all those worried about the problem of freedom and authority in our time. He sees freedom as creativity for the good, not as liberty to do as one pleases.

Fourth, although this may be debatable, Gregory's perspective of Christianity makes possible a more open attitude toward those outside the Christian fold—beyond that of converting them and bringing them inside the fold. Gregory can relate the incarnation of Jesus Christ to the whole of humanity and to the whole of creation, not just to Christians.

Fifth, for those who are now weary of our three centuries old cosmology—this so-called scientific, secular, modern cosmology—

Gregory delivers us from enslavement to any kind of deification of modern rationality and its twin product of science and technology, while at the same time maintaining their values. We do not have to oppose divine wisdom to modern knowledge, but we can put the latter in its place. It is time we reexamined and reassessed such concepts as enlightenment, modernism, rationality, and so forth. And Gregory helps us do so in a way that few other Christian thinkers to date can. Indeed, the study of Gregory of Nyssa helped the present author to evaluate, and ultimately to transcend the above concepts. If there is any value in existentialism, phenomenology, process philosophy, analytical philosophy or structuralism, Gregory's framework allows us to integrate them in a coherent manner.

Sixth, it is the author's conviction that modern Christian thinking needs a perspective like Gregory's in order to overcome the superficial quality it has developed. Gregory's framework makes it possible to reinterpret the incarnation of Jesus Christ in a way that should eventually be acceptable to Catholics, Protestants, and Orthodox alike, and that makes better sense to non-Christians. His perspective is neither narrow nor sectarian. It is universal and humanist in the best sense.

No wonder Maximus the Confessor (580–662 A.D.), who was Secretary of State to Emperor Heraclian (610–641) of the Graeco-Roman Empire before he became a monk, called Gregory the "Universal Teacher" or "Teacher of the *Oikumente*." In 787 A.D., Second Council of Nicea (called Seventh Ecumenical) spoke of Gregory as "named by everybody as Father of Fathers" (*ho pater pateron*).[1]

Nearer to our time, we can cite two of the greatest students of Gregory in the last few decades. The comments of the late Cardinal Danielou, the doyen of Gregory scholars are given in the introduction to the first edition:

> The work of Gregory . . . combines the toughness of research with loyalty to the faith. His work is in touch with the thought of his time, but it is not enslaved by it. It conveys together the meaning of being and the meaning of history. It

1. Mansi, *Amplissima Collection* (Paris–Leipzig, 1902). Tom. xiii, col. 293.

> brings together confidence in the capacity of the mind to apprehend reality, and the sense of inescapable mystery that surrounds everything that the mind so apprehends. These are the things that truly respond to the questions we are asking today."[2]

Hans Urs von Balthasar, who calls Gregory "the most profound Greek philosopher of the Christian era, an incomparable mystic and poet"[3] also speaks of him as surpassing Origen and Gregory Nazianzen in the profundity of his thought, and concludes:

> There is in Gregory's Christian detachment some sort of naive and ardent attachment to all the beauties of the earth, for all the human and worldly values, which gives to his thought a tone of youth and the freshness of dawn."[4]

For Balthasar the three Christian fathers to study were Origen, Nyssa, and Maximus. Of these he thought Nyssa to be both the most profound and most philosophically attractive.

Western scholarship has generally been unkind to Gregory, probably because he did not fit the Augustinian, Thomistic or Reformation categories into which some scholars tried to put him. Others have dismissed him as an unoriginal platonist. Such critics have generally been philologists with no interest in Christian theology, and have judged him in terms of literary style. Gregory was neither a stylist like his brother Basil, nor a poet like his colleague Gregory Nazianzen.

Some of these scholars were also impeded, at the time of their writing, by the lack of a proper critical edition of Gregory's works. This has been remedied. There is now a special institute for Gregory of Nyssa research at the University of Muenster, West Germany, where the present writer did a good part of his research. There are at least three concordances to Gregory's works now in preparation at universities in Europe and America. There has been in existence for some time now an International Society for Gregory of Nyssa

2. J. Danielou, *L'etre et le temps* (Leiden, 1970), p. x.

3. H. Urs von Balthasar, *Presence et Pensée* (Paris, 1942), p. xiii. Balthasar's magnificent work in French is, alas, out of print. Both a new edition and an English translation seem to be called for.

4. H. Urs von Balthasar, p. xiv.

Studies, which brings together the few dozen western scholars now engaged in Gregory studies.

Gregory of Nyssa was as much a philosopher as he was a theologian. The present work does not attempt to summarize his thought. The attempt here is to sketch the main tenets of Gregory's understanding of the three fundamental realities in their mutual relationship—i.e., God, humanity, and the created order.

Paulos Mar Gregorios
New Delhi

INTRODUCTION

This book seeks to analyze the problem of human existence between its two poles, God and the Creation, in relation to the writings of one 4th century Christian thinker, Gregory of Nyssa, born around 330 A.D. and deceased around 395 A.D.

The first part of the work is epistemological and the second part ontological. The epistemological discussion starts with the background of the 4th century theological debate between Eunomius and the Cappadocians. It then lifts up two major epistemological principles in Gregory's thought—*epinoia* and *akolouthia*, in relation to the Christian Scriptures.

The second part, for the sake of convenience, deals mainly with the dialectic between *diastema*/discontinuity and *metousia*/continuity. The key concept of *pleroma* is also then analyzed, in relation to sinful human existence in history.

The conclusion pulls together the main structure of Gregory's thought and relates it to problems posed by our contemporary civilization.

The method used is basically to adduce texts from the writer which reveal his thinking on certain particular aspects. In view of the extreme length of some of the very significant texts, they have had to be relegated to the appendix.

The book also seeks to show that the attempts to classify Gregory as a Platonist or a Neo-platonist cannot stand examination. Neither can his thought be understood as some sort of 'mystical theology'. This is a category developed in the later history of the West and cannot be adequate for apprehending Nyssa's thought. It also seeks to dispel the notion that Gregory was unoriginal and was merely slavishly or at random borrowing from other thinkers. It deals with the background of certain ideas in order to show that Gregory uses the ideas of other thinkers only after examination by a consistent category—the scriptural Revelation combined with the Principle of *Akolouthia* or coherence.

The book does not shy away from pointing out inconsistencies or from being highly critical of some of Gregory's ideas. On the whole, however, the effort is to present the thought of Gregory of Nyssa as a valid alternative and necessary corrective to current or historical ways of western theologizing. The author confesses a certain bias against western theology. This however, should not be any more an impediment to understanding Gregory than would be the opposite bias which uses Augustinian or scholastic categories or a modern liberal understanding of freedom as the standard for evaluating the thought of Gregory.

1. The Theme and Its Relevance

The theme of this book focusses on humanity's two basic relationships—to the source and ground of its being on the one hand, and to the created world in which humanity is placed on the other. It is the contention of this work that in an authentic Christian understanding of reality like that of Gregory of Nyssa these two relationships are inseparable from each other. There is no Christian way of understanding reality in a totally secular sense, as if Man and the World were the only two realities with which we have to deal. But then it is equally disastrous for a Christian to think that we can conceive of a God who is concerned only about our souls and has no relationships to the Creation. Neither of our two basic relationships can be conceived, from a Christian perspective, except in terms of the other.

It would be fatal for us, therefore to go on oscillating between, on the one hand, an other-worldly mysticism that ignores the reality and significance of humanity's sinful existence in history, and on the other, a secular humanism that ignores the ground and source of the being of ourselves and the cosmos. Humanity itself becomes distorted and prone to self-destruction when it ignores either pole. Christianity, if it is to be alive in this time, should find a proper perspective of reality that takes both poles adequately into account. No authentic Christian anthropology can be conceived except in the framework which locates man as existing dynamically between these two poles. This work does not seek to provide a full anthropology of Gregory of Nyssa or to deal exhaustively with Nyssa's Christology, soteriology or theology. It simply provides the structural framework within which the other aspects of Gregory's teaching can be fittingly placed.

The structural frame-work is so decisive for our perception of reality, that slight distortion in the framework can lead to signifinant error of judgement in living and shaping lives in the world. Starting from a wrong structural framework we may device very successful sciences and technologies, but in the end the whole thing must come to grief because of the underlying structural error.

To speak more concretely, scientists and philosophers of science are today raising basic structural questions about man's relationship to his world. Is nature inclusive of man or exclusive of man? That is one form in which the question is put by ecological thinkers today.

All civilizations have at their base a cosmological-anthropological structural perspective, which they often take for granted. Our contemporary civilization also has one such—a curious and novel structural perspective, as Edwin A. Burtt sees it in his *The Metaphysical Foundations of Modern Science:*

> "How curious, after all, is the way in which we moderns think about our world! And it is all so novel, too. The cosmology underlying our mental processes is but three centuries old—a mere infant in the history of thought—and yet we cling to it with the same embarrassed zeal with which a young father fondles his new-born baby. Like him, we are ignorant enough of its precise nature, like him we nevertheless take it piously to be ours and allow it a subtly pervasive and unhindered control over our thinking"[1]

Professor Burtt tells us that "the central metaphysical contrast between medieval and modern thought" is that in medieval thought man's place was more central than that of nature, while in modern thought "nature holds a more independent, more determinative, and more permanent place than man". The modern world is thus born through a displacement of man from the centre of the structural perspective, the vacancy being filled by an "objective" nature which, de-deified and secularized, becomes open to our science and our technology. The visionary poet Dante and the philosopher-scientist Bertrand Russell stand as the symbols of the two ages and their cosmologies. Dante's *Divine Comedy* yields place to Russell's human tragedy of an "objective world"

1. Revised Edition. Doubleday Anchor, 1932. p. 15.

where, standing "on the firm foundation of unyielding despair", modern man can worship only "at the shrine that his own hands have built".

But today we are beginning to tire of that heroic stand. Alienation has so caught up with us that we dread this shrine which our own hands have built; for it may at any time collapse bringing the roof down over our heads. It is time that we looked again at the structure of the shrine of this "modern" world in which we have worshipped for these last two or three centuries.

As the cosmological perspective changes, our language also usually undergoes a thorough overhaul. Words like substance, accidents, causality, essence and idea yield to new ones like forces, laws, fields, movements and energies. These latter too are not entirely new. They are often taken from prevailing systems of the day or from forgotten systems of an earlier day. If Newton for example could assume that the world was composed of bodies moving about in space under the influence of definite and measurable forces, he took the idea over from a prevailing ideology of metaphysical mathematics, as Burtt shows. This ideology, as well as Newton's own, had dealt with three realities—God, Man and World. For Newton too, God was the super-governor of the universe, who originally created bodies with mass and set them in motion. He also keeps it going.

In later Deism, it was assumed that though God may have set them going, once they got going, they functioned through the laws that God had written into them, without any need for further intervention from God. Once Newtonian theism can thus lead to Leibnizian Deism, the next stage of transition to secularism is understandable, where a God-reference is no longer necessary to explain the origin of the universe, to guarantee its continuing existence.

If that is in some sense how we came to our "modern world" where we are the subject while the world is the object and there is no "third party" to worry about, we should once again look at our structural perspective. Burtt, after examining our modern structural perspective, comes to the interesting conclusion that we can have an adequate cosmology only when we have an adequate philosophy of mind, one—

> "that provides full satisfaction both for the motives of the behaviourists who wish to make mind material for experimental

> manipulation and exact measurement, and for the motives of idealists who wish to see the startling difference between a universe without mind and a universe organized into a living and sensitive unity through mind, properly accounted for."[2]

Burtt is modest enough to admit that the whole thing is beyond him. He recommends, however; "as an indispensable part of its foundation" a "clear historical insight into the antecedents of our present thought-world". It is as a part of the search for such an insight that this work is offered.

Unfortunately however, Nyssa offers no solution that would satisfy Burtt. Nyssa's world knew strict logicians like Eunomius who could handle academic technology with great skill and speculative idealists like the school of Plotinus who could see mind or *nous* as the true reality of all that exists, but the experimental method in relation to the human mind was not highly developed in his time and the questions of the behaviourists had not yet risen. We have therefore to be satisfied with but a cursory and finally unsatisfactory sketch of Gregory's epistemology. The focus will have to be on his ontology. The epistemological preoccupation will have to be subordinated to the ontological one. The question that matters here is: "what is humanity, placed between God and His Creation?" The other question is also important: "How do we know?". But it will have to be treated as a subsidiary question.

But one may question, from the perspective of science, this tendency to relegate epistemology to a subsidiary position. Science, so wedded to precise measurement, cannot function without procedures to test the validity of the instruments used to measure. If mind is the instrument by which the reality of man's place between God and the universe is to be known or measured, then we must first check and make sure that the mind is capable of knowing it and knowing it accurately. This is the modern perspective from which Burtt insists that we need first to have a satisfactory philosophy of mind before we can attempt an adequate cosmology.

This again is a characteristic of our "modern" world. We seek certainty about any knowledge by tracing the sources and method

2. Burtt. *op. cit.* pp. 324-325.

of that knowledge. Knowing how we know is necessary in order to know that what we think we know is really knowledge. This characteristic quest of our modern world is best exemplified in Descartes, who much more than Newton, seems to be the father of our modern world. In his *Meditations on the First Philosophy* (1641) Descartes perfects his method, but his quest is still for the knowledge of these three realities (God, Man and World), though he does not focus on their mutual relationship in any systematic way. In his meditation I, Descartes lays down his starting point, the foundations of the certainty of the system, as the clear and indubitable fact of one's own mental activity.

Descartes is here in basic continuity with the hellenic tradition, especially the Second Academy, which used reason to question the certainty of the self-evident. When all that is known by the mind is put in question, only one thing remains unquestioned, or at least so the Western tradition thought ever since Augustine, namely the indubitability of the doubting mind—the *res cogitans.* All experience has to find its explanation from that single starting point, the thinking mind. The consciousness of the perceiving ego is the foundation rock upon which all modern intellectual systems of certainty stand. Intellectual systems, including that of modern science presuppose these three elements:—the consciousness of the perceiving ego, the particularity of the experienced object, and the purity of the reasoning process. But first always is the thing that thinks—*res cogitans.*

Descartes, educated by the Jesuits at La Fleche, is here bringing back one of the basic methodological principles of the great father of Western spirituality—Augustine of Hippo. This indubitability of Man's soul as "conscious nature" or "thinking essence" was Augustine's sheet anchor which rescued him from sinking in the sea of skeptic doubt. The bishop of Hippo tells us in both the *Enchiridion*[3] and the *Retractations* that he was in these early days trying "to raise up an unbreakable dam against the destructive floods of skepticism"[4]. And he too scans the whole field of knowledge to locate

3. *Enchiridion* 20 "One cannot err who is not alive. That we live is therefore not only true, but it is altogether certain as well" (Eng. Tr. Albert. C. Outler in Library of Christian Classics: Vol. VII· *Augustine: Confessions and Enchiridion* London, 1945. p. 322). See also *Contra Academicos (Answer to Skeptics)*: 2:9:23.

4. A.C. Outler, *op. cit.* Intro. p. 89.

a few (not one, as in the case of Descartes) indubitable *truths*. For Descartes the one single indubitable truth is to serve as a starting point for mathematically systematic reflection. For Augustine the main concern is to show that total skepticism is both unnecessary and dishonest. Later Thomist interpreters [5] of Augustine have tried to show that the future bishop of Hippo was actually giving expression, though without using precise terminology, to the scholastic logic of Aristotelian metaphysics—the three principles of identity, non-contradiction and excluded middle. Augustine himself did not name any principles, but spoke of some *a priori* truths of logic, which are *rationes aeternae,* somewhat similar to eternally valid logical principles. And it is from these logical principles of permanent validity that one derives one's basic certainty.

But Augustine did not regard reason in itself as either self-sufficient or inerrant. In his *Contra Academicos* written at the age of 33, he concludes that one has to be impelled by the dual authority of faith and reason, and not by reason alone, in order to arrive at the possession of truth. But clearly it is still the consciousness of man that is the base and locus of certainty. In his *de Trinitate* (written in his late fifties or early sixties) he still sticks to that line, but the youthful expression of *Contra Academicos* is more unequivocal:

> "Certainly, no one doubts that we are impelled toward knowledge by a twofold force: the force of authority and the force of reason. And I am resolved never to deviate in the least from the authority of Christ, for I find none more powerful. But as to what is attainable by acute and accurate reasoning, such is my state of mind that I am impatient to grasp what truth is—to grasp it not by belief, but also by comprehension. Meanwhile, I am confident that I shall find among the Platonists what is not in opposition to our sacred scriptures".[6]

5. e.g. Denis J. Kavanagh, in Intro to *Answer to Skeptics* in series *The Fathers of the Church, Writings of St. Augustine*, Vol. I. New York, 1948. pp. 91-92.

6. *Contra Academicos*: 43. Eng. Tr. Denis Kavanagh. *op. cit.* p. 220.

In our own time, it is important to recognize once again with Augustine that a pure skepticism and an easy agnosticism are no longer luxuries we can afford. We are under pressure to overcome our fear of ontology and metaphysics, and launch a fresh attempt to perceive reality in its rich depths

Not that such ontological or metaphysical reflection can be expected to appeal immediately to the common public. Discussion on certain basic questions may not interest the public primarily because it lacks the training and background for following the discussion. But when the issue is clarified and the consequences are worked out in actual life, then the public begins to see the importance of the original discussion.

The issue we have in hand is that of the structural perspective of the cosmology that undergirds our civilization. We examine the thought of St. Gregory of Nyssa because his thought is an alternative to certain ways of cosmological thinking that have become natural to scientists or theologians. Our own cosmology will have to be much more sophisticated than Gregory's, because we know much more about the structure of the universe and of matter and energy, of cells and life, and of sub-atomic particles than Gregory could know in his time. But the way he has shown is well worth examining.

2. The Man Gregory and His Place in the Tradition of the Church

Gregory of Nyssa, younger brother of St. Basil the Great, was born probably between 335 A.D. and 340 A.D. The details of his biography have been admirably investigated by Gerhard May in his contribution to the "Chevtogne Colloquium"[7] on Gregory of Nyssa studies.

The father, the elder Basil, was a Rhetorician, who must have taught a great deal of classical philosophy to his own sons and daughters.[8] Gregory shows wide and detailed knowledge of the dialogues of Plato, the dialectic of Aristotle, and the Stoicism of

7. *Die Chronologie des Lebens und der Werke des Gregor von Nyssa* in M. Harl, ed. *Ecriture et Culture* (Chevtogne Colloquium Report), Leiden 1971, p. 53 ff.

8. See Gregory Nazianzen's *Homily on Basil* xii 1-2 Migne's *Patrologia Graeca* (PG) 36:509 A-B.

Poseidonius. He was also well acquainted with the Enneads of Plotinus and the writings of Porphyry. Gregory had a special interest in the science of his time; he knew his anatomy as best as he could in his day, probably with the aid of Caesarius, the brother of Gregory Nazianzen. Caesarius was Court physician to the Emperor, and left the Imperial palace to join the monastic community of the Pontic woods started by Basil and the Nazianzen.

Gregory was certainly more enthusiastic than his brother Basil in his attitude towards non-Christian literature and knowledge. Though he had less formal schooling than either of the other two Cappadocians, he had read more philosophy than both Basil and Gregory Nazianzen. Gregory calls himself "uninitiate in logic"[9] and yet an ardent student of the second sophistique like Louis Meridier[10] can argue that the whole work of Gregory of Nyssa is a conscious imitation of the logical pyrotechnics of this unrestrained Asiatic literary school which began in the first century of our era.

This knowledge of logical technology was consciously acquired by Gregory of Nyssa partly for reasons of showiness, but more properly because it was necessary to combat the purveyors of theology who infested the streets, especially after Julian the Apostate came to power and pagan philosophy began to reassert itself.[11] Gregory became so well-versed in the technique and arguments of the pagan rhetoricians that his sister, the saintly Makrina had to chide him about his over-attachment to pagan wisdom and its ways[12].

9. *amuētos tou logou*, Epist. XIII: 4, GNO. Vol. VIII: 2. p. 45 PG 46:1049A.

10. L. Meridier, *L'influence de la seconde sophistique sur l'oeuvre de Gregoire de Nysse*, Paris, 1906.

11. In that biting satirical sermon preached at Constantinople entitled *On the Deity of the Son and the Spirit*, Gregory breaks forth against these street-vendors of theology: "The whole city is full of this kind of people, the narrow lanes, the markets, the public squares, the dual highways—tailors and money-changers, vendors of groceries. If you ask one for some coins of change, he will philosophize to you about begotten and unbegotten. If you ask one for the price of a loaf of bread, he will reply, 'the Father is greater, and the Son is subordinate'. Supposing you say you want a bath, the bath attendant will say, 'the son's being is from non-being'. I don't know what name to give to this evil, this mental perversion, craziness or madness, plague or epidemic". PG 46:557B.

12. *de Anima et Resurrectione* PG 46:20A. Eng. Tr. Nicene and Post-Nicene Fathers (NPNF) Vol. V: 431B.

In fact Gregory was so attracted to pagan wisdom that for a while he seems to have left the service of the Church and become a Rhetorician. If Gregory Nazianzen was shocked and grieved about this 'secularization' of a Church Reader[13] (in 364 A.D.?), he also anticipates Nyssa's possible sarcastic response that by going among the young intelligent students as a Rhetorician he would be better able to practice his Christianity. Basil and the Nazianzen believed that bishops should keep away from too much contamination through pagan contacts. Nyssa on the other hand seemed to have believed, at least in this early period, in the necessity of direct contact with unbelieving intellectuals. Nyssa is grateful for and respects pagan knowledge even while making fun of it, while Basil and the Nazianzen are a bit paternalistic in their evaluation of the intellectual achievements of unbelievers.

Basil, of course, looked down upon all—whether it be the arianizing Emperor in Constantinople, the Prefect Modestus, the Bishop of Rome, or even his brother the bishop of Nyssa. The towering personality of the big brother kept Gregory in the shadow, and it was only after Basil's death on January 1st 379 that Gregory's star began to rise. The mantle of Basil fell upon Gregory, and though the Nazianzen was still living, it was upon Gregory that the leadership of the Nicean Party devolved.

St Basil's death may have been hastened by his fairly hot temperament. Basil quarrelled with everybody—including cousin Nazianzen and brother Nyssa, with arianizing Emperor Valens and with the staff of the Emperor's household. Everybody was afraid of him while respecting his integrity and wisdom. Only Eunomius, the leader of the Arian movement, could be sarcastic and scurrilously personal in his attacks on Basil. It is quite likely that the publication (in 378?) of Eunomius' reply (*Apology for an Apology*) to Basil's attack on him hastened on the latter's death.[14]

But this controvery with Eunomius' strict-sounding logical analysis was what shaped Gregory's basic intellectual structure, as we shall see.

---

13. Greg. Naz. *Epist. XI*. PG 37:41B-44B. The authenticity of the address of this letter has been disputed. On this see P. Gallay, *Les Manuscrits des lettres de S. Gregoire de Nazianze*. Paris 1957. pp. 44-45.

14. See the allusions in Gregory's *Contra Eunomium*: Bk I, written around 380 A.D. PG 44:61-124. GNO I.

Gregory has had a fairly bad press in the history of Western theology. His intellectual achievement has been consistently played down, and it may not be unfair to suspect that by the time Gregorian writings became sufficiently well known in the West, the Augustinian-Thomist system of Western thought had struck deep roots in Western intellectual soil, and the radically different Gregorian framework was to most medieval and post-medieval theologians largely unacceptable. It was in the 9th century, through the Latin translations of John Scot Eriugena (ca 810-ca 877) that Gregory became known in the West.

The East had recognized Gregory's genius much earlier. Maximus the Confessor (580-662 A.D.) had praised him as the "Ecumenical Teacher"[15] and owed much of his framework to Gregory. The second Council of Nicea (called Seventh Ecumenical—787 A.D.) referred to Gregory as "named by everyone as the Father of Fathers".[16]

In our own century it was Hans Urs von Balthasar that drew our attention to him with his superbly poetical work *Presence et Pensée.*[17] His enthusiasm was kindled by the praise bestowed on Gregory by Prof. Wilamowitz and Prof. Werner Jaeger. Cardinal Danielou who takes the credit for the first more balanced western study of Gregory in our time[18] called him first the 'greatest mystic among the Greeks' and only later learned to give a more balanced evaluation:

> "The work of Gregory. . . combines the toughness of research with loyalty to the faith. His work is in touch with the thought of his time, but it is not enslaved by it. It conveys at the same time the meaning of being and the meaning of history. It brings together confidence in the capacity of the mind to apprehend reality, and the sense of inescapable mystery that surrounds everything that the mind so appre-

---

15. Maximus. *Opuscula et Polemica Theologica,* PG 91:161A. Gregory was recognized by Imperial decree as official teacher of the Imperial Church: *Codex Theodosianus* I:XVI. tit. 1.1.3. Cited by Hefele 'Leclercq' *Histoire des Conciles* tom I: 1. p. 41,

16. See Mansi, *Amplissima Collectio*. Paris, Leipzig, 1902. Tom. 13. col. 293.

17. Paris, 1942.

18. *Platonisme et Theologie Mystique,* p. 6.

hends. These are the things that truly respond to the questions we are asking today".[19]

That succinctly sums up our own interest in Gregory's thought. That thought is not scholastic to the extent of eliminating the element of mystery; but then neither is it an unintellectual mysticism. It seeks to do justice both to the questions of fundamental ontology and to those about the meaning of existence in history. It is thought which is old as the rocks and fresh as the dawn. It soars into the transcendent without losing sight of the material and the historical. It plays fair with pagan thought without being disloyal to the scriptures. Surveying the whole intellectual field of the 4th century Byzantine culture it manages to escape being dated, and speaks with fresh relevance to the issues of our time.

19. *L'etre et le temps*, Leiden, 1970. Intro. p. X.

Part One

SKOPOS—AGENNESIA—AKOLOUTHIA

CHAPTER I

SKOPOS
Intent of Scripture and the Use of Outside Philosophy

There are two charges usually made against Gregory's use of his sources—uncritical adoption of Platonic thought[1] and unauthorized use of the allegorical method in the interpretation of Scripture[2]. We should examine quickly to what extent these two charges are justified, and in the process learn something about his special way of using Holy Scripture and pagan knowledge.

1. The Intent of Scripture

The matter of Gregory's use of Scripture was extensively studied by Gregory scholars in the first Gregory Colloquium held at Chevetogne, Belgium in 1969. One thing Monique Alexandre showed clearly in her paper on *The Theory of Exegesis in 'On the Making of Man' and in the Hexaemeron*[3]. The difference between St. Basil and St. Gregory is not that the former uses the literal method, while the latter uses a 'spiritual' interpretation, as Cardinal Danielou tried to establish. Basil wrote for the ordinary man, Gregory wrote for the intellectual. Gregory has therefore a concern for philosophical coherence which is not very evident in Basil. The younger brother has also a concern to reconcile what appear to be mutual contradictions in Scripture, like for example, the two accounts of

1. See for example the work of Harold Cherniss, *The Platonism of Gregory of Nyssa, Berkeley*, 1930.

2. e.g. H. Weiss, *Die Grosser Kappadozier Baselius, Gregor von Nazianz und Gregor von Nyssa als Exegeten*, Braunsberg, 1872.

3. See Marguerite Hall (ed.) *Actes du colloque de Chevetogne, Ecriture et Culture Philosophique dans la Pensee de Gregoire de Nysse*, Leiden, Brill, 1971, 267 pp. See pp. 87-110.

creation in the first chapters of Genesis. Gregory's concern is to resolve these contradictions within an adequate philosopical framework. Hence his concern for a fresh interpretation of Scripture that by resolving contradictions illuminates reality.

Holy Scripture is a way opened by God for the mind to direct itself to the God who created it. Scripture opens our eyes to see the design of God[4] and thus leads us to God's mind and purpose. But the meaning of Scripture itself is not always self-evident. Certain passages in Scripture, says Gregory, are like peacock's feathers. The side that is first visible may be dull grey, but if you turn it around its beauty and glory are manifested. Such turning around can be done only for one who is grounded in the faith of the Church. Otherwise he stands in danger of mis-interpreting the Scripture. Every passage of Scripture has to be interpreted in consonance with the faith of the Church. This is one sense of the word *akolouthia* which is a central category for Gregory. The faith of the Church has consequences for our understanding of the meaning of scriptural passages as well as for the evaluation of "outside teaching" which is what Gregory calls the knowledge and wisdom of the pagan schools. The principle is most clearly stated at the beginning of the *Refutatio Confessionis Eunomii*[5].

Gregory speaks of the faith of the Church as of divine origin, and as the light that guides to the truth, in our understanding of Scripture as well as in our sifting of outside knowledge. This does not mean however, that local custom can be used to contradict Scripture. The teaching of the Church has to "agree with divine words"[6]. Holy Scripture is God-inspired or *theopneustos*, but this does not mean that every individual by his own free will can understand the meaning of Scripture. The true intention or *skopos* of Scripture is evident only to one who lives in the faith of the Church—especially the teaching about the Holy Trinity and the teaching about the Incarnation of Jesus Christ. This means that there is no one given method for interpreting every passage of Scripture—whether it be literal, spiritual, typological or allegorical. For one who lives in the faith of the Church, the *skopos* of Scripture becomes clear and he would know which method to use to interpret

4. See *de anima et resurrectione*, P.G. 46. 442B. also *In Hexaemeron* PG 44:69D.

5. See Appendix I.

6. *Ad Eustathium:* GNO III:1.5/20 to 6/6 NPNF V:327A.

a particular passage, so that its meaning does not contradict the faith of the Church. This became clear to Gregory in his dispute with the heretic Eunomius, as we show in the following chapter.

Here we need to deal with Gregory's application of the category of the *skopos* or intent of Scripture to evaluate pagan philosophical insights. Gregory clearly does not believe that a rational philosophy by itself can yield true knowledge. The dialogue between Gregory and his sister Macrina is the *locus classicus* for this view. Gregory plays the role of the defender of the syllogism as a source of truth, but says that "for us, however, we confess the teaching demonstrated by the Holy Scriptures to be much more reliable than all the conclusions of technical logic"[7].

Macrina chides her brother for putting logic and Scripture on the same level, and insists that the Scripture alone is fully reliable. In the long speech of Macrina, Gregory rejects the view that reason is sufficient proof of truth. Reason, in its speculative thrust, is free to go wherever it wants, whereas for the Christian, reason has to be controlled by the category of the *Intent of Scripture.* Gregory says that the pagan thinker can use reason to affirm what he wishes, while for Christians,

> "We make the Holy Scripture the standard and rule of all teaching; we are bound, therefore, to have in view that and only that which is in harmony with the intention of Scripture"[8].

Macrina therefore says that Christians have to abandon the Platonic understanding of the soul (as a chariot, with two dissimilar horses, reason and passion, pulling it in different directions); Christians have to reject the Aristotelian understanding of the soul, which through logical-technical analysis comes to the conclusion that the soul is mortal:

> "We must reject all those before these as well as after these, whether they philosophized in plain prose or rhythmic verse. We must make the scope of our reasoning the God-inspired Scripture (*skopon ton logon tēn theopneuston Graphen poiēsō-*

7. *de anima et Resurrectione.* PG 45:64 A-B. NPNF V:442B.
8. *Ibid*, see PG 46:49 C. NPNF V: 439 A-B.

> *metha*) which regards the soul as having no other property than that which participates in the divine nature. For he who says that the soul is in the likeness of God, admits also that all that is alien to God is outside the limits of the nature of the soul. Dissimilarity cannot be defended in what is similar"[9]

Thus it is on the basis of the faith of the Church, which regards the Scriptural notion of creation in the image of God as central to the understanding of humanity, that Gregory can reject the anthropology of Plato, Aristotle and the Stoics. In all major matters, Scripture or rather the intent of Scripture, understood in consonance with the intent of the faith of the Church becomes the category by which to judge Plato, Aristotle or the Stoics—whose views dominated the "outside philosophy". There was no attempt on the part of Gregory to accommodate the insights of the gospel to suit the preferences of prevailing philosophies. On the contrary, his adherence to the Church's understanding of reality was strong enough to enable him to reject the prestigious views of Plato, Aristotle and the Stoics.

We can illustrate the same point by looking at what Gregory did to two other prevailing assumptions of outside philosophy—one that Man is a *microcosm* of the universe, and the other that the whole universe breathed together as a single organism—the concept of *sympnoia*. Before we do that, however, it needs to be pointed out that Gregory's use of "outside" wisdom was not confined to philosophy alone. He opened himself to all branches of culture, literature, music, art, medicine, engineering, and biology. He listened to contemporary rhetoricians like Libanios and Themistios, as well as to scientists like Theon of Alexandria, who taught from 378-395 A.D., developing the mathematics of Euclid and Ptolemy. Gregory probably followed the thought of Oribasius (ca 320-400 A.D.), the personal physician of Emperor Julian the Apostate and author of 70 or 72 medical works, incorporating all previous Western medical knowledge.[10] Gregory frequently visited medical doctors, and acquired an immense amount of anatomical knowledge, conceivably with the help of Caesarius, the brother of Gregory Nazianzen, who

9. *Ibid.* PG 46:49B to 52 A, NPNF. V: 439 A-B.

10. Oribasius' *Collectiones Medicae (Iatrikai synagogai)* have survived only in part—about 25 works. Through Syriac and Arabic translations of Oribasius, Greek medicine *(yunani)* entered the Islamic world.

had been court physician before Oribasius and who later joined the Pontic monastery of Basil[11].

Gregory knew something also about art, music and architecture. The sculptor's art is described in the commentary on the Psalms;[12] he gives detailed instructions on Church architecture to Amphilocius.[13]

His knowledge of music shows up in the *Commentary on the Inscriptions of the Psalms;* where the whole creation becomes one giant musical instrument, which when properly tuned, bursts forth in the celestial music of a cosmic symphony, our own hymnody and psalmody being participations in this cosmic music.[14]

Thus Gregory exemplifies a positive attitude to the material creation as such, and not merely to pagan philosophy. The mistrust of pagan knowledge, art and culture, of beauty and pleasure, which characterised certain aspects of the early Latin tradition are not seen here. It is a fact that the better values in pagan culture were more freely appreciated and faithfully conserved by Gregory than they were by many contemporary pagans, who under the influence of Neoplatonism and Manicheeism despised the aesthetic values of this world.

In order to appreciate this basically positive approach of Gregory to the material creation, we have to observe two apparently contradictory facts about his thought. On the one hand he attacks the pagans who regard human science, language and poetry as divine creations. He would de-divinize science and language over against an Eunomius who regarded the human mind as practically worthless and regarded divine names as well as the names of things as handed down by God himself. For Gregory it does not demean

11. Caesarius died in 368. Jose Janini Cuesta who has studied Gregory's medical knowledge, thinks the latter inherited Caesarius' medical *library*. See J.J. Cuesta, *La antropologia* y *la medicina pastoral de san Gregorio de Nisa.* Madrid 1946. Gregory expressly mentions Galen and Hippocrates (PG 45:164 A) and the last chapter of *de hominis opificio* gives a summary of the physiological knowledge of Gregory's time (PG 44: 245 D, NPNF V: 422 B ff. For the description of a doctor's diagnostic technique see PG 46:29 C-D. NPNF V: 434 A)

12. *In Inscrip. Psalmorum:* XI GNO V:114/25ff PG 44:541ff.

13. See this remarkable epistle XXV in GNO VIII:II pp. 79ff. PG 46:1093ff NPNF. V:Letter XVI. p. 540ff.

See also Gregory's description of fresco painting with specifications in the homily on the Martyr Theodore (de S. Theodoro Martyre. PG 46:737. D-740A).

14. *In Inscrip. Psalmorum III.* GNO I:30/15ff PG 44:440Bff.

language to say that it is a creation of man, nor does it detract from the glory or the grace of God, for all that man has that is good is a reflection of that glory and a gift of that grace. In the pagan controversy whether language has its origin along with the *phusis* created by God, or by the *thesis* or ascription of man, Gregory takes definitely the latter side,[15] as does St. Basil (but not Origen). It is no disgrace for language and science to be a human creation.

On the other hand, Gregory has a higher evaluation of human capacity than Eunomius and his Neoplatonist school. The human faculty of knowing and creating is a noble thing, not a thing to be despised, though it becomes despicable when it exceeds its natural limits and capacities.

The fourth century was an age very much like ours—an age of prosperity and affluence when Philosophy becomes devalued and science-technology gains the upper hand.[16] Gregory had the unusual ability to create a philosophical system which neither was antagonistic to science nor failed to make use of it. But his intellectual system was not a mere integration of the philosophic and scientific traditions. He had sufficient faith to make the Christian tradition a touch-stone for the evaluation and appropriation of philosophical and scientific insights and discoveries. He integrated science and philosophy on the foundation of the Christian tradition. This principle of integrating science and philosophy on the basis of the Nicean tradition of Christianity, using Trinity-Incarnation as central category is what we hope to illustrate, by taking two key terms of Gregory's contemporary science and philosophy—the concept of *sympnoia* of the cosmos and the idea that man is a *microcosmos*.

It is not sufficient, as many experts in source research (*Quellenforschung*) have done, to point to the parallelism of the occurrences in pagans and Christians in the use of certain terms. One has to see carefully how these terms are used, what place each term holds

15. The supposed words of Democritus of Abdera (fl 420 B.C.) that "words were statues in sound, but statues not made by the hands of men, but by the gods of themselves" (see NPNF vol. V, p. 269B note) do not appear among the fragments of Democritus cited by Diels. What we do have is the reference in Proclus' commentary on Plato's Cratylus (423B). Gregory's convincing argument against Eunomius is in GNO I. 277ff (PG 45. 969ff, NPNF V. 268 A ff.).

16. Danielou recognizes this in his *IVème Siècle*, p. 35.

in either system, and how they are basically transformed in the process of re-integration into a system of which the central affirmations came from the Christian tradition.

2. Sympnoia

Stoic ontology begins with the assumption that there are four elements which constitute the universe—air, water, fire, and earth. But what holds these four elements together? Later stoicism advanced the hypothesis that there is a fifth element which pervades all things and holds the universe together. This they called *pneuma* or spirit. Galen, the physician (130 - ca.200 A.D.) attributes the origin of this idea in Medicine to Athenaios (first half of first century A.D.), the founder of the Pneumatic School in medicine:

> "According to Athenaios, the elements of man are not the four primary bodies, fire, air, water and earth, but their qualities, the warm, cold, dry, and humid... He introduces a fifth element in accordance with the Stoics, the all pervading pneuma by which all things are held together and controlled."[17]

Clement of Alexandria agrees with the idea of the fifth element, which he calls "the tension of the pneuma which pervades and holds the cosmos together,"[18] but he regards this pneuma as identical with aether as was suggested by Empedocles.

While Sextus Empiricus (fl ca. 180 A.D.) would attribute the cohesion to a force called *sumpatheia* and not to a fifth element,[19] and Plutarch (ca. 46 - ca.127 A.D.) attacked the Stoic idea of the fifth element from a Platonic standpoint, the most determined attack on the idea of the fifth element came from Alexander Aphrodisiensis (became head of the Lyceum in Athens towards the end of the second century A.D.) who as official commentator on Aristotle felt obliged to point out inconsistencies in the idea of the fifth element.[20]

---

17. Galen, *Introductio S. Medici*. 9. Engl. Tr. in Sambursky, *The Physics of the Stoics,* Appendix, p. 119.

18. *Stromateis* V. 8.

19. Sextus was a Cynic who directed his attacks against Stoic cosmology and epistemology. See his Adv. Math. IX. 78-80 in R.G. Bury, ed., *Sextus Empiricus*, Loeb, 4 vols. 1939-1957.

20. See especially *De mixtione*, p. 216. 14ff. and 223. 6ff.

I. Bruns, ed., Alexander Aphrodisiensis, *Scripta minora*, 2 vols., Berlin 1887 and 1892.

From these criticisms it becomes clear that the theory originated in Chrysippus (ca. 280-207 B.C.) himself, convert to Stoicism and successor of Cleanthes as head of the Stoa from 232 B.C.[21] Chrysippus seems to have been some sort of a St. Thomas in the history of Stoicism. He produced a tightly coherent systematic cosmology in which the universe is one single concatenation where nothing happens without being caused by something else and every cause leads by necessity to its appropriate effect. The concept of *Eimarmene* or fate thus becomes not a simple fatalism, but a manner in which man can know the cosmos and gain mastery over it through knowledge and vision.

Gregory appropriates this Stoic insight in its basic form and integrates it with the Christian doctrine of creation, thereby departing radically from Stoic cosmology. The words *sumpneō* and *sumpnoia* are used by Gregory, but only rarely,[22] and in most cases the reference is to the human body and its parts rather than to the cosmos. Most of the instances cited by Cardinal Danielou in his chapter on *Conspiration* in *L'être et le Temps chez S. Grégoire de Nysse* relate to the human body and not to the cosmos, and he rightly notes that the word was a medical term.[23] It is only in the catechetical oration and in *Contra Fatum* that Gregory refers to *sumpnoia* in a cosmological sense, and even in these instances he is only referring to the pagan use of the term without necessarily approving it.

This reluctance of Gregory to use *sumpnoia* in a cosmological sense is best accounted for by the association of the word with *heimarmenē* or fate in Stoic thought. In his treatise *Against Fate,* Gregory makes clear that for the Stoics Fate is a kind of God who indwells the universe and activates it (*theos tis estin tō onomati tes heimarmenēs*).[24] He explains the Stoic belief thus:

"Since the *sympatheia* in all that exists is one, and since the

21. Chrysippus spoke of the *he ton ouraion pros ta epigeia sumpnoia*—Arnim, *St. Vet. Fragmenta* 2. 172.

22. Only three times in *Contra Eunomium* viz. vol. I. 25/17, I. 86/6 and II. 258/25. Only once in *In Inscriptio Psalmorum* (V. 32/12), only once in *In Ecclesiastes* (V. 406/4) and only twice in *De Virginitate* (VIII. 1. 292/4; 322/21).

Not all these usages have the context of the whole universe, but often relate to the human body, or to the Body of Christ.

23. *op. cit.*, p. 51.

24. PG 45. 152 A.

whole universe is continuous within itself, and that which is understood separately is to be comprehended as in one body with one spirit (*epi somatos henos en mia sumpnoia katalambanetai*), all the parts converging upon each other, they regard the upper region as the most primary principle to which all the things around the earth are coordinated."

There can be no doubt, reading through the account, that Gregory has some original Stoic work before him as he writes. He gives an ample account of the theory which he then proceeds to refute. It is quite possible that what he has before him is a medical man's account of Stoicism, for it is the consequences of this cosmology for medicine that Gregory narrates in some detail.

But the spiritual consequence of this Stoic cosmology seemed unacceptable to Gregory. For what it amounts to is that man is a plaything of this great mighty primal force called Fate; for his life being linked inescapably with everything else, the movement of the stars affects his movements. He becomes merely a small pawn in the giant cosmic chess-game. This explains Gregory's reluctance to accept *sympnoia* in its Stoic sense. The concept of a fate which guides stars and human beings alike Gregory categorically rejects. He rejects also the Chaldean cosmology which had already penetrated Hellenism, especially through Porphyry and which made the stars and the planets the controllers of the universe, because they belong to the upper regions, and the earthly is subordinated to the heavenly movements.

Danielou fails to note this negative element in Gregory's attitude towards the notion of *sumpnoia*. While Gregory does accept the basic idea of an inter-connectedness for all the elements of the universe, he does not use the term *sumpnoia* for that inter-connectedness. Danielou correctly notes the Platonic elements in Gregory's conception,[25] but he fails again to note that Gregory practically gives the term an ecclesiological meaning in contrast with its cosmological use by the Stoics. It is significant that in his Epistle 17, Gregory uses *sumpnoia* for the whole Church acting together as a single body. Cardinal Danielou sees this primarily as denoting

25. *L'être et le Temps*, p. 55, but he gives little evidence for his statement: "*Mais la théorie particuliere de Grégoire releve de la tradition platonicienne.*" (ibid)

simply "unity of will,"[26] but it is obviously the ontological unity of the Body of Christ that Gregory has in mind. The problem in the Church in Nicomedea is that it is in discord; the presbyters who rule the Church in the absence of a bishop are not agreeing with each other. The *homonoētikē sumpnoia* which Gregory demands is a characteristic of the Body of Christ, and the *sumpnoia* of the Church is dependent not merely on the various parts breathing together, but also on their being in symphony with the will of God, preferring it to their own.[27] Thus the *sumpnoia* of the Church is a transcendent one, and not of the type that the Stoics conceived the *sumpnoia* of the cosmos to be.

Danielou notes this ecclesiological orientation of the concept *sumpnoia* in Gregory in connection with other passages in Gregory,[28] and agrees that it is necessary, according to Gregory, to belong to the Body of Christ in order to participate in Christ's *sympnoia*, or the power of Christ's life. Gregory can claim that even if he is weak, as a participant member of the Church, he is stronger than Eunomius, since the weakest part of the Body of Christ is stronger than that which has been (like Eunomius) separated from that Body.[29] But in this instance Gregory does not use the term *sympnoia* as, for example, he does more explicitly in *de Perfectione* or in the commentary *On the Beatitudes*, or even more fully in the commentary *On Ecclesiastes*, from which last we cite the relevant passages apropos Ecclesiastes 3:6b—which says "a time for keeping, a time for throwing away" (Jerusalem Bible).

> "The following-up of this passage leads the soul to a greater philosophy of existents (*eis meizona tina tēn peri tōn ontōn philosophian*), namely that the universe is continuous in itself, and there is no rupture of the harmony of existents; rather there is some sort of a co-breathing (*sumpnoia*) of all among each other, neither can the whole universe separate itself from this mutual conjunction, but all things remain existent only when activated by the power of Him who truly is. . . That

---

26. *op. cit.*, p. 58.

27. GNO VIII. 11 (Ep xvii) pp. 57/22—58/4. PG 46. 1057 B ff.

28. *op. cit.*, p. 59. See *Beat.* 8. PG 44. 1293 B. *De Perfectione.* GNO VIII. I. 197/19-24. PG 46. 213 A. *C.E.* I. 10. GNO I. 25/17. PG 45. 252 A.

29. *C.E.* I. 10. See note 28.

> which is outside Being has no being. Now evil is opposed to the good (*aretē*). But God is full perfect good. So then evil is outside God, and has thus no being in itself."[30]

If this is so, the presence of evil is a kind of rupture within the chain of being. So long as evil exists in the world and in us, the concept of *sumpnoia* cannot be rigorously applied to the creation, but only to the community that has been redeemed from the *akolouthia* of evil. Here we see Gregory separating himself from Stoic doctrine in another way. The lack of *sumpnoia* is due to sin or evil, whether it is in the Church or in the cosmos. True unity, abiding unity, is not possible except in the good. This is Gregory's firm view, which makes it difficult for him to accept the Stoic doctrine of cosmic *sumpnoia* in its Stoic sense. The ontological cannot be separated from the *kalological*, to create a neologism for the systematic understanding of the good.

Danielou adduces several neoplatonic parallels for a supposedly similar use of *sumpnoia* in Plotinus, Iamblichus and Proclus. But they do not have the same rigorous identifications of *to on* and *to kalon*. Gregory has again learned both from the Stoics and from Middle or New Platonists. But he has made their doctrine fundamentally different by relating the two concepts of being and good closely to each other. It is the Cross of Christ that now extends in all four directions to drive out evil, and to bring everything back again not merely to unity, but more specifically to unity in the good.[31] Not merely unity in the good as such, but as comprehended by Christ, through the Cross, through a united Church. Christ was extended on the Cross, in order he may occupy all the dimensions, the height and the depth, the length and the breadth of the whole creation, as Gregory understands St. Paul to have said in Ephesians 3:15. This theme of unifying through the Cross the world which lost its unity through the entry of evil is a recurrent one in Gregory.[32]

---

30. *In Ecclesiasten, Oratio VII.* GNO V, pp. 406-407. PG 44. 724D—725B.

31. *Oratio Catech.* 32. PG 45. 80-81. See also *De Tridui spatio.* GNO IX. 1, pp. 299-300. PG 46. 621-624.

32. See e.g. *Oratio Catech.* 32. *De Tridui spatio.* loc. cit. *Contra Eunomium* III. 3. 39-40. GNO II. 122-123. Full discussion in G.B. Ladner, *St. Gregory of Nyssa and St. Augustine on the Symbolism of the Cross* in *Late Classical and Medieval Studies in honour of A.M. Frend, Sr.* Princeton University Press, 1955.

But this unity is not yet fully realized. It is to be realized only on the last day. The Cross is a symbol of this eschatological unity.

Here one sees how radically un-Stoic and how deeply Christian Gregory's thought is. While on the one hand he accepts the basic notion of *sumpnoia* from the Stoics, he radically christianizes it,

(a) by denying that there is true *sumpnoia* in a creation which has been invaded by sin;

(b) by centring the *sumpnoia* on Christ and his Cross; and

(c) by making the *sumpnoia* eschatological.

This is a typical illustration of how Gregory uses pagan knowledge as filtered through the Christian tradition and integrated with the biblical perspective on the universe.

It is in this context that one recognizes the wisdom of Danielou's words:

> "The study of all the texts has convinced us that there was no need to investigate what were the Platonic elements in the thought of Gregory, but we should accustom ourselves to a view of his thought which is purely Christian, but which is expressed in the philosophical categories of the time in which it was constituted. . . The Platonic structure is here only accidental. It is substantially the same kind of thought which could express itself in Buddhist technical categories, but would not for that reason be any less Christian mystical thought than Buddhist mystical thought. . . One can say, in general, that Gregory has allegorized everything, even philosophy. The Platonic language, especially that of the myths, offers him, along with the Bible, a treasury of expressions speaking for those who care to listen, of the unique mystery of the transfiguration of the soul in Jesus Christ."[33]

3. Cosmos and Microcosmos

The idea that man is a universe in miniature, a *microcosmos*, goes back to the pre-Socratics in Greek thought. The word *mikrokosmos* does not occur in Lidell and Scott's Greek-English Lexicon,

33. Danielou, *Platonisme et Theologie Mystique*, p. 9. (Engl. Tr. present author).

neither does one find it in Gregory. But the idea is most clearly expressed in the commentary *On the Inscriptions of the Psalms:*

> "We have heard one of the wise men discoursing about the nature of our soul, that man is some kind of a small universe (*mikros tis kosmos estin ho anthropos*), having in himself all the elements of the large universe."[34]

and again;

> "If the arrangement of the whole (universe) is some kind of a musical harmony, whose technician and maker is God, as the Apostle says, and if man is a small-scale universe, (*mikros de kosmos ho anthropos*), then man himself is also a copy of the harmonious universe; even as we saw the work of reason in the larger universe, we see the same in the small one. For the part of the whole is homogeneous in everyway with the whole. As for example in even the smallest fragments of a glass, on each polished surface, the whole orb of the sun can be reflected as in a mirror, however small the polished surface, so also in the small universe, I mean in human nature, the whole music of the universe can be discerned, the whole being proportionately reflected in each part."[35]

Here the argument is that man's singing psalms to the glory of God is proportionate to the whole cosmos singing the praises of God. But does this mean that Gregory simply accepts the *whole* of the Stoic doctrine that man is a microcosmos of the universe?

The idea of microcosmos, in its historical development from Anaximander to Paracelsus has been described at length by Rudolf Allers,[36] though the historical treatment, admittedly incomplete, leaves too many lacunae to be satisfactory. Unfortunately, Allers

---

34. *In Inscr. Psalm.* I. III. GNO V. 30/24ff. PG 44. 440.

35. *Ibid.* GNO V. 32. 16-33. 1. PG 44. 441 C-D.

36. R. Allers, *Microcosmus, From Anaximandres to Paracelsus* in. J. Quasten and S. Kuttner, ed., *Traditio*, vol. II (1944) pp. 319-407.

See also H. Lotze, *Mikrokosmus* (place and date not known) and G.P. Conger, *Theories of Macrocosms and Microcosms in the history of Philosophy*, New York, 1922.

does not document the contention that Anaximander (ca. 560 B.C.) actually used this idea.

It seems credible that microcosmism had its origin in the Pythagorean world, and later found a central position both in the medical terminology of Hippocrates and Anaximander, and in the cosmology-anthropology of the Stoics. If we want to locate specific texts, one could find a fragment of Empedocles of Acragas (ca. 450 B.C.) who had seen Love as the element that holds the diversity of the cosmos from going apart, and which also holds man together,[37] both the cosmos and man being formed out of the same four elements. As for the words *macrocosmos* and *microcosmos*, it seems the atomists came closest.[38] The titles of Democritus' works as given by Diels includes a volume on Physics with two parts—1. *megas diakosmos* and 2. *mikros diakosmos*.

Plato certainly became a source for much later speculation about this proportionality between cosmos and man. In *Philebus*, Plato argued that the orderly movement of the cosmos could not be explained by chance, but had to be attributed to *nous*. As our bodies are regulated by a rational soul, so is the cosmos directed by superhuman souls.[39] *Timaeus*[40] goes farther and became the *locus classicus* for later commentators on microcosmism, especially in the neoplatonist schools. The world-soul is there clearly stated to be the archetype of the human soul. Neoplatonism transformed Plato's notion of proportionality between world-soul and human soul, into the doctrine that the human soul was derived from, or an emanation from, the world-soul. One sees the idea of world-soul and the astral bodies as animated by *nous* and *psuchē* criticized and rejected by Philo Judaeus.[41] He uses the terminology of a world-soul but only in the sense that the Logos is immanent in the world; but the Logos is also transcendent.[42] But for Philo also

"if the part (man) is an image of the image (i.e. of the cosmos),

37. Diels, pp. 229ff. E.T. Kathleen Freeman, *Ancilla to Pre-socratic Philosophers*. Fr. 17 on p. 53, Fr. 20 on p. 54, Fr. 22 on p. 55 etc.

38. *Vierte Auflage, zweiter Band*, Berlin 1922, p. 19, line 27.

"Great World Order", "Small World Order". See Freeman, *Ancilla*, p. 92.

39. *Philebus* 28d to 30c.

40. *Timaeus* 34b ff.

41. Wolfson, *Philo* I, pp. 346-347.

42. See e.g. Philo, *De Vita Mosis* II. 25. 127.

> and if the complete image, the whole of the sensible world, is something greater than the human, and is a copy of the divine image, it is clear that the archetypal seal, which we call the intelligible world, is the Logos himself."[43]

Gregory feels uncertain about this analogy between the cosmos and man, whether it has any real place in a Christian way of thinking. The words he puts into the mouth of his sister Macrina would give the impression that Gregory finds the idea generally acceptable.

> "Says she: It is said by the wise that man is some sort of a microcosm (*mikros tis einai kosmos ho anthropos*) encompassing in himself these elements by which the universe is constituted and perfected. If this view is true, (and so it seems), perhaps etc. . ."[44]

Even in these words there is no enthusiasm for the idea that man is a microcosm. Elsewhere Gregory rejects this view as clearly misleading for it gives occasion to the thought that man's unique dignity is derived from his resembling the cosmos rather than being the image of God. In *On the Making of Man*, Gregory says:

> "How petty and how unworthy of the greatness of man sound the imaginings of some outside philosophers who presume to magnify man by comparing him to this cosmos! For they say that man is a little cosmos (*mikron einai kosmos ton anthropon*) being constituted of the same elements as the universe! Those who with such a high sounding name seek to bestow praise on human nature seem to be oblivious to the fact that they are thereby dignifying man with the attributes of the gnat and the mouse! For they too are a mixture of the same four elements, since each single existing being, whether great or small, we behold only as participating in a portion of these sensible elements, otherwise they would have no nature. What is so great then, in man's being reckoned an impression and likeness of this world—of the heaven that passes away, of the

43. *De opificio Mundi*. 25.

44. *De Anima et Resurrectione*. PG 46. 28B. NPNF V. 433A.

earth that becomes alienated, or of any of the things contained in it, since they too are to pass away with that which contains them?

But what then is the greatness of man, according to the teaching of the Church? Certaintly not in his likeness to the created cosmos, but in his coming to be in the image of the Creator's nature."[45]

We see clearly Gregory rejecting the pagan concept of man as microcosm precisely because it seems to compete if not conflict with the basic Christian affirmation that man is created in the image of God. Yet he is not averse to using the notion of microcosmism in a limited sense, as we see in the first part of *In Inscriptiones Psalmorum* where he refers to the idea in a more positive way:

"We have heard one of the wise ones say in a passing comment on our nature, that man is some kind of microcosm (*mikros tis kosmos esṭin ho anthropos*), having in himself all the elements of the large cosmos. For the world-order of the universe is some sort of a musical harmony (*harmonia tis esti mousikē*), harmonized with itself according to a certain order and rhythm intricately interweaving much diversity, in unison with itself, and never discordant with this symphony, despite the immense range of difference between each of the various existent beings."[46]

Gregory goes on then to describe how the heavens tell the glory of God, and how the harmony of the creation raises a symphony of music to the glory of God. Gregory is prepared to accept the pagan idea of man as microcosm only in this limited and definitely Judaeo-Christian sense, namely that as the cosmos continuously lifts up a hymn of praise to God, so it is the duty of man, the smaller cosmos, to engage in continual psalmody and hymody.

Gregory thus rejects the pagan idea that man is to be defined as microcosmos, i.e. that what distinguishes him is the fact that the logos that brings order to the world by controlling it harmonizes man also by indwelling his body. One difficulty for this idea, both

45. *De opificio hominis*. XVI. PG 44. 177 D—180 A. NPNF V. 404 A—B.
46. GNO V. p. 30/24 to p. 31/1. PG 44. 440 C.

for Platonists and Neoplatonists would be their tendency to regard the soul as a stranger to the body, imprisoned in it, which is not of course the situation of the logos in the cosmos. But Gregory rejects it not in terms of its logical inconsistency alone. He wants to maintain the central insight of the Christian tradition that man is made in the image of God. In the light of this category Gregory finds pagan anthropological microcosmism unsatisfactory. Yet he makes use of it in a limited sense, namely to assert that as the cosmos is a musical symphony for the glory of God man also engages continually in praising the glory of God and joining the cosmic symphony of God's glory.

It should become clear from the examples given that Gregory is no slavish eclectic in matters of philosophy. Was Gregory a Platonist? A Neoplatonist? An Aristotelian? A Stoic? All or any of these labels would not fit Gregory.

Take Plato for example. Philo was definitely a Platonist. Gregory was not. About Philo Prof. Wolfson says:

> "The starting point of Philo's philosophy is the theory of ideas. This theory was with him a philosophic heritage from Plato, and according to his own belief, as we have seen, also from Judaism."[47]

The theory of ideas is in a sense distinctive of Platonism, and if it is also central to Philo's system, there is some justification in calling Philo a Platonist. But one seeks in vain in Gregory of Nyssa to find a theory of ideas, not to speak about its being central, or a starting point. We shall discuss in greater detail in a later chapter how Gregory transforms the Platonic theory of ideas in his doctrine of creation.

But there is a certain similarity of approach between Philo and Gregory.

> "Like all of his followers (i.e. the Christian, Moslem and Jewish philosophers of a later age), he (Philo) also started with the belief that there could be no real contradiction between Scripture and Philosophy. Like all of them, therefore, he must have been aware—we have reason to assume—of the

47. Wolfson, *Philo* I, p. 200.

> fact that certain contradictions do seem to exist between Scripture and philosophy, and that these contradictions would have to be removed. We have also reason to assume that he was not less perceptive than they in seeing that certain philosophic views were absolutely irreconcilable with the teachings of Scripture. Similarly we have reason to assume that he was not less ingenious than they in knowing how some refractory philosophic views, with certain revisions, could be reconciled with scriptural teachings. So also we have reason to assume that he was not less painstaking than they in examining thoroughly every philosophic view before deciding whether to accept it or not.........If this is how we are to approach the study of Philo, then to get at the true meaning of his philosophy it is not sufficient to collect related passages in his writings, to arrange them under certain headings, and to place in their juxtaposition parallel passages from other philosophers and the Bible. We must try to reconstruct the latent processes of his reasonings, of which his uttered words, we may assume, are only the conclusions."[48]

What Professor Wolfson says about Philo applies even more eminently to Gregory. Gregory does not reveal the whole process by which he arrives at certain conclusions in relation to pagan philosophy, but his conclusions point to a rather systematic process for examining pagan ideas, and for assimilating, rejecting or revising them in a way of thinking which derives its basic inspiration from the Christian faith.

One does not overlook here the fact that a competent scholar like Werner Jaeger feels that Gregory is a Platonist, since the essentials of Platonism which Gregory has chosen for his thought are the true essentials of Platonism.[49] We need to examine this statement of Jaeger in the light of Prof. Endre von Ivanka's analysis in *Plato Christianus*.[50] Von Ivanka argues that there are two mutually separable ideas in Platonism, one, the theory of ideas, i.e. that things can be known, not in their concrete reality, but only in their pre-existent

48. Wolfson, *Philo* I, pp. 104-106.
49. W. Jaeger. *Two Rediscovered Works*, p. 138.
50. Einsiedeln 1964, esp. see pp. 151-185.
See also *Vom Platonismus zur Theorie der Mystik* in *Scholastik* 1936, pp. 163 ff.

forms or ideas of which all particulars are imperfect manifestations, and two, the teaching about the inner knowledge of God, i.e. that we do not know the Absolute God through any one of the particular existents, but by a progressive delving deeper into the ground of our own inner being.

Gregory definitely rejects the world of ideas, and at this point departs also from Neoplatonism which re-incorporated Plato's world of ideas into its second level of being—the nous which is an emanation from the One. For Neoplatonism only the One is transcendent, as it is for Eunomius also. But Gregory sees how incompatible with the Christian doctrine of Trinity-Incarnation these doctrines of the transcendence of the One, the emanation of the intelligible world from the One, and the very world of ideas are. And he rejects both the Platonic and the Neoplatonic interpretation of the world of ideas.

Von Ivanka argues that Neoplatonism had already accomplished the separation between the two ideas in Plato, i.e. between the knowledge of particular existents through their ideas or forms and the knowledge of the One by descending into oneself, and seeing the One with the eye of the soul and not through the existent particulars. In this latter way of "knowing", Plotinian neoplatonism held that the soul has, in order to attain to the knowledge of the One, to divest itself of all knowledge of particulars, and even of its own self, so that, thus purified, the soul "sees" the One in the ground of its own being. Plotinus says at the conclusion of the Enneads:

> "The main difficulty is that this awareness *(sunesis)* of the One comes neither through knowing *(kat' epistēmēn)* nor through intellection *(kata noēsin)* as in the case of other intelligibles, but by a presence surpassing knowledge *(kata parousian epistēmēs kreittona)*. The soul is subject to an apostasis or separation from being One, and is no longer One, in the process of taking knowledge. For knowledge is account or discourse *(logos gar hē epistēmē)* and account or discourse is multiple *(polla de ho logos)*. The One thus strays into number and falls into multiplicity. So we must proceed beyond knowledge and never depart from being One. We must leave aside knowledge and the objects of knowledge, and from all else, even those that seem good. For this that we seek is

> beyond all good, and all good is derived from this, even as all light of day is from the sun. For this reason it is said 'Neither spoken nor written'. But we speak and we write only in order to send (people) to it. From words we would lift them up to vision, as if showing the way to those who wish to have vision. Our teaching is only about the road and the journey, the vision must be the work of him who has desired to see."[51]

Does Gregory basically accept this "outside philosophy"—this second part of Platonism as developed by Plotinus, with the help, as we have suggested earlier, of Upanishadic thought mediated to his teacher Ammonius Saccas or to himself, by Hindu thinkers? Von Ivanka says that the view of Plotinus at this point is no longer strictly Platonic, having been already separated from the key to Platonic thought, namely the doctrine of ideas.[52] It is, according to von Ivanka, however, this Neoplationic idea of the knowledge of God that Gregory finally accepts. Von Ivanka points out many parallels in Gregory—in the *Homily on Gregory Thaumaturgus*,[53] in *de Virginitate*.[54]

The parallels cited by von Ivanka are certainly impressive. On the first reading, it would appear that he has made his case that it is Plotinian thought that Gregory follows. We need not doubt that Gregory both has read Plotinus and was deeply influenced by him. But does he transform the thought of Plotinus substantially in the process of integrating it into his system of Christian thinking? Von Ivanka admits that the recurrence of this theme in Gregory is not to be attributed to a direct dependence on Plotinus, but rather to the fact that Plotinian images had already entered the Christian tradition and appear even in Gregory Thaumaturgus as well as in Clement of Alexandria.[55] He also admits that in Gregory of Nyssa, the soul does not lose its alterity and identity as creation, which is

---

51. *Enneads* VI. 9.4. Bréhier's edition. Tom. VI, p. 176.

52. *Plato Christianus*, p. 160.

53. PG 46 : 908c *en eschatiae tini heautǭe monǭe sunēn, kai di' heautou toē theoē.*

54. PG 46 : 368D *to di' autēs dunēthēnai ton theon idein.*

Also PG 46 : 416D (end of the book) *hina kai autos idēis ton theon en katharae tēi kardiae.*

55. *Plato Christianus*, p. 163 note.

a point of radical difference from Plotinus in whom the soul is not created but only an "ecstasis" or standing apart from the One. Von Ivanka recognizes that Gregory speaks only of the soul returning to its original state as created and not of becoming identical with God, and also that for Gregory even this is not the result of human efforts, but the grace of God in Christ. Von Ivanka can also see that the doctrine of grace as it developed in the west has its roots in Aristotelian ontology and epistemology,[56] and that the idea of enclosing man within a clearly defined "nature of man", so that what is "super-natural" falls outside the nature of man, would be unacceptable to Gregory or to other Eastern Fathers.

But one regrets to say that the learned professor von Ivanka seems caught in the categories of western thought and seeks to force Gregory of Nyssa into a Platonic-Neoplatonic mould, though he makes several qualifications which taken together, would nullify his main thesis which seeks to see Gregory basically as a Platonist and a Neoplatonist.

Gregory of Nyssa rejects almost all pagan affirmations which he found contrary to Scripture and the Christian tradition. He rejects Plato, Aristotle, the Stoics and Plotinus in his views about the soul, about resurrection, about the nature of the world, about the nature of man, and about God. These are the central realities of Christian thought, and here Gregory is neither Platonist nor Neoplatonist.

He accepts their terminology—this is quite natural if one wants to communicate with cultured pagans, and to defend the Church against the philosophical onslaughts of pagans and philosophical heretics like Eunomius and the Macedonians.

Gregory has radically altered even the neoplatonic notion of knowing God in the depth of oneself, through his relating it to the Christian idea of the Image of God. What one sees in the depths of the purified soul is not a direct vision of the incomprehensible God, but a proportionate reflection, as in a small mirror, of the Infinite and the Eternal. We shall discuss this in detail in a later chapter. But here it needs to be pointed out that no basic idea of pagan philosophy has been accepted by Gregory without transforming it in the process of integrating it into a Christian prespective on reality.

56. *Ibid.*, p. 173.

Ekkehard Mühlenberg, in a scholarly article on *Die philosophische Bildung Gregors von Nyssa in den Büchern Contra Eunomium*, argues that Gregory's philosophical formation can be considered separately from his biblical-Christian formation, since in his controversy with Eunomius, the debate is often on a purely philosophical basis, with Scripture playing only a sub-ordinate role.[57] Mühlenberg tries to reconstruct the philosophical systems of Eunomius and Gregory, to set them over against each other, and then to draw conclusions about the philosophical antecedents of Gregory's system.

But the methodology used by Mühlenberg is certainly questionable. The assumption that Scripture plays only a sub-ordinate role in Gregory's mind (not merely in his writing) needs to be proved. For Gregory's basic view of pagan philosophy is that it can be used for or against the truth. As Eunomius uses it against the truth, Gregory feels constrained to use it in favour of the truth, but it is not his philosophical formation which tells him what is the truth to be defended, but his training in the Christian Church. Gregory's philosophical arguments should not be examined as if he were objectively and impartially seeking answers to questions through philosophical discussion. He knows already that Eunomius' argument that Unbegottenness is the essence of God, and therefore that only the Father is fully God, is mistaken. That it is mistaken Gregory does not discover from philosophy, but from the Christian Church. His philosophical training helps him to find the philosophical arguments by which to refute the philosophical arguments of the heretic. The attempt "to reconstruct the library of Gregory" is useless, for that library contained everything he had been able to get hold of, but that does not mean that all the books there gave him his philosophical category-structure.

Gregory knows Aristotle's categories, Plato's ideas, Plotinus' Enneads, Stoic science, and the works of Epicurus and Porphyry, Panaetios and Poseidonios, but the attempt to separate his philosophical reasoning from his theological reasoning is possible only for a western thinker. As David Balás pointed out at Chevtogne, "these 'philosophical' analyses are to be found at the interior of a structure and argumentation which are 'theological'."[58] Mr. van Parys who has made a detailed study of Gregory's exegetical method, also

57. In M. Harl, ed., *Ecriture et Culture Philosophique*, pp. 230-251.
58. M. Harl, *op. cit.*, p. 244.

questioned[59] Mühlenberg's statement that Scripture and theology play only a subsidiary role in the Eunomian debate. And as Gerhard May pointed out, if we take that debate on a purely philosophical ground, it is difficult to give the victory to Gregory. Eunomius' system is philosophically neat, though absolutely wrong, as a Christian has to say.

I would conclude this chapter by drawing the attention of the reader to the letter of Origen to Gregory's spiritual ancestor, Gregory Thaumaturgus, which sets forth the right attitude of a Christian towards "outside philosophy". We do not know whether Gregory of Nyssa actually read this letter of Origen. But the attitude there presented, namely that pagan philosophy can be either the handmaid of Christian teaching, or the occasion for idolatry in the Temple of God, seems very close to Gregory's own view.

Gregory would also agree with Origen, that it is only rarely that "outside philosophy" becomes truly a means of glorifying God. In Gregory's case, perhaps more than in the case of Origen, the wisdom of the Greeks and the gold of the Egyptians was used for the true glorification of God.

59. *Ibid.*, p. 245.

CHAPTER II

AGENNESIA—AGAINST EUNOMIUS

The Eunomian Controversy as Background for the Clarification of Gregory's Thought-structure

It was the controversy with Eunomius that set the stage for the clarification of Gregory's Christian thought. The history of this controversy can now be reconstructed from the somewhat scanty sources available to us. Prof. Dr. Mrs. L. Abramowski of Bonn University has most ably summarized the main points about Eunomius and his thought in the *Reallexikon für Antike und Christentum*.[1] We need only to qualify her conclusions slightly. The only two known full studies of the controversy remain unpublished.[2]

Eunomius was a contemporary and chief adversary of the Cappadocian Fathers. He was also a Cappadocian, from Oltiseris in Cappadocia on the border of Galatia. Gregory of Nyssa had to justify the mistake by Basil in calling him a "Galatian."[3] Gregory gives us some details of his life, perhaps not quite objectively. Son of a farmer, Eunomius learned shorthand (*tachygraphia* as it was called), went to Alexandria and became disciple and secretary[4] of the arch-heretic Aetius, the sophist disciple of Arius himself. From Aetius, Eunomius learned the technology (*technologia*) of sophis-

1. Stuttgart, under *Eunomios,* Cols 936-947.

2. (a) Th. Dams, *La Controverse Eunomeenne*, Typescript 294+58 pages of notes, without date or place.

(b) Bernard C. Barmann, *A Christian Debate of the Fourth Century: A Critique of Classical Metaphysics* unpublished Ph.D. Dissertation, Stanford University 1966 (not consulted)

3. *Contra Eunomium.*

4. Socrates, *Hist. Eccl.* IV. 7. *Contra Eunomium.* I GNO I, pp. 38ff. PG 45. 264 ff. NPNF vol. V, pp. 40ff.

tic reasoning. Eudoxius of Antioch,[5] himself a semi-Arian, ordained him to the deaconate, and later when Eudoxius was promoted to be Archbishop of Constantinople, he ordained Eunomius to be Bishop of Cyzicus (360 A.D.). Eunomius had an impressive capacity for the display of sophist dialectic, and was able to make out quite a name for himself. But his people soon found out the emptiness of his life. Besides he quarralled with Eudoxius himself by insisting that Aetius be re-instated.

Within a year, he was forced by his people to resign his episcopate and he left for Cappadocia. When Julian the Apostate became Emperor (361-363), the Anhomoian[6] party of the Arians reorganized themselves under the leadership of Aetius and Eunomius, fighting both the Orthodox (led by the Cappadocians) and the Homoian party of Arians (led by Eudoxius of Constantinople and Euzoius of Antioch).

Around 363, when Julian died, Eunomius began organizing a Church of his own, and ordained bishops of his party in various sees. Around 366/367, with the death of Aetius, Eunomius became the unquestioned leader of the Anhomoian party, and moved to Constantinople as his headquarters, only to be exiled soon by Emperor Theodosius.

The importance of the Eunomian crisis is seldom recognized by Church historians. It was as much a peril as the Arian crisis two generations earlier with which the battle was joined at Nicea by St. Athanasius. But precisely because the Athanasian-Nicean settlement did not deal adequately with the philosophical problems involved in Arianism, it continued to survive among intellectuals and ordinary people alike. Aetius, with his sharp logic learned from the second sophistic, spread the doctrine far and wide. There was a new form of liberal "Christianity" developing which denied the very foundations of the Gospel, but was eminently acceptable because of its conformity with current philosophical trends.

5. Socrates calls him an opportunist and a fraud. *Hist. Eccl.* II. 38.

6. The Arians, who generally refused to acknowledge the full deity of Christ, were divided into two parties, both of which refused the Nicene formula that Christ was *homoousios* (of the same *ousia*) with the Father. The Homoian party of Eudoxius was willing to accept the formula that Christ was *homoiousios* (of like nature) with the Father. The *Anhomoian* party refused even this formula, and declared that Christ was *heteroousios* (of a different nature) from the Father.

Aetius and Eunomius were seeking to indigenize Christianity by domesticating it within the current and acceptable philosophical framework which we have already referred to as Alexandrian philosophy. Basil, Gregory Nazianzen, and Nyssa, equally or even better at home in current philosophy, saw it clearly that this kind of indigenization would destroy the Gospel itself which was sought to be indigenized. Hence it is of the utmost importance, say for example in India, to see the two different approaches to indigenization, one followed by Aetius and Eunomius, and the other by the Cappadocian Fathers.

What was the teaching of Eunomius? All the great Orthodox thinkers of the fourth century—Didymus the Blind,[7] Basil the Great,[8] Gregory of Nyssa,[9] and the more doubtful but equally powerful thinkers like Apollinaris of Laodicea,[10] and Theodore of Mopsuestia,[11] as well as a certain Sophronius of whom we know little, found it necessary to compose treatises against Eunomius. It is from these criticisms that we are able to reconstruct the Epistemology and Ontology of Eunomius, which provide the background for understanding the Epistemology and Ontology of Gregory of Nyssa.

What was the essence of the difference between the two attempts to indigenize theology, the one by the Cappadocians representing Nicene Orthodoxy, and the other by Eunomius, coming in the Arian-Aetian tradition? *Eunomius sought to adopt Christianity to secular philosophy. The Cappadocians sought to adopt secular philosophy to the Christian Gospel.*

That is a fundamental difference, extremely relevant to the problems of theology today. What is the central criterion which is the cornerstone of our way of theological thinking? Does it come from

---

7. The two books of Didymus *De dogmatibus* and *Contra Arianos* mentioned by Jerome in his *De viribus illustribus* 109, have been identified, by some with the 4th and 5th books appended to St. Basil's *Contra Eunomium* in many MSS and printed editions (PG 29. 671-744). See Quasten, *Patrology*, vol. III, p. 88.

8. Basil's *Contra Eunomium* is in PG 29. 497-669.

9. Gregory's major work is *Contra Eunomium,* in Jaeger's Edition volumes I and II.

10. Apollinaris' work is not extant. See Quasten, *Patrology*, vol. III, p. 378.

11. Theodore's most famous work: *De Incarnatione* is directed against Arius, Eunomius and Apollinaris, See Quasten, *op. cit.*, p, 410.

the Christian faith or from secular philosophy? This question is basic.

For the Cappadocians the basic category is Trinity-Incarnation. For Eunomius the foundation is the unity, simplicity and absolute uniqueness of the One—a category that reigned supreme in the Alexandrian "secular" philosophical milieu.

In the Alexandrian philosophy there is no room for a Trinity, a distinction of persons, within the one Godhead. Neither is there room for God becoming Man. Only a created being could become Man, according to Eunomius. That is the only way he could find to integrate the doctrine of the Incarnation within the prevailing philosophical system.

Dams comes to the following conclusion in his dissertation:

> "So there is no more need to restate that the problem of knowledge is the crucial point at which the two theologies confront each other. We have seen that in Eunomius it forms the base of his system, while with the Cappadocians it is equally at the base of their speculations; but with this qualification, however, as we have just said, that the faith should be one principal source, and perhaps even the first subjective principle, for religious knowledge......Gregory and Basil, as well as Eunomius, proceed in an eclectical fashion as regards philosophy; but with the Cappadocians faith operates with so vital a force as to give all their reflections a profound unity, in fact a personal system, a rule of life."[12]

Eunomius himself states that there are two ways of doing theology—one to study the natures *(ousiai)* of Father, Son, and Holy Spirit in themselves; the other to study the energies or operations *(energeias)* of the natures. Eunomius chooses the first; Gregory chooses the second.

What is the methodology that Eunomius uses to study and compare the natures of the Father and the Son? Certainly not *epinoia*, the capacity of the human mind to conceive realities.

The essential nature of God is unbegottenness or *agennēsia.* The name *agennētos*, applied to God, is not simply a human creation. It denotes the very essence of God, that which distinguishes him

12. *op. cit.*, p. 294.

from all that has come to be—*to gennēma*. For what the human mind creates is expressed as an utterance—a sound that vibrates in the air—and not reality itself. But the name *agennētos* is not a human creation. It is the revealed name of God, expressing his *ousia* as the source and cause of all that has come to be.

Now no one can say that the Son, the Only-begotten (*monogenēs*) is unbegotten. Thus there is complete difference of *ousia* between the Father, who is *agennētos* and the Son, who is *monogenēs*.

These two names of the Father and the Son, according to Eunomius, are not created by the human faculty of *epinoia*, but given by revelation in the Scriptures and confirmed by the Fathers. Thus the revealed names show the basic dissimilarity in essence between the Father and the Son. There is no community of essence (*koinōnia tēs ousias*) between the Father and the Son.

Eunomius would not thus admit that he is yielding to the demands of pagan philosophy about the unity and simplicity of the One God. He would rather insist that his teaching about the *heteroousia* of the Father and the Son is squarely based on the names given to them by the Scriptures and confirmed by the Fathers. In fact, Eunomius is himself fighting against heresy, particularly Sabellianism which made the Father and the Son merely two modes of the same person.

Eunomius' epistemology is an attempt to find the true sense behind the revealed names—the *physikē ennoia*, the intuition of the essence, terminology that reminds one of modern phenomenology. The names correspond perfectly with the reality—for Eunomius—as the *noēma* corresponds perfectly with the external object in Husserl's phenomenology.

In contrast to Gregory's basic dichotomy between Creator and creation, Eunomius proposes another dichotomy—*agennētos—gennēma* between the Unbegotten and that which is begotten. In Gregory's dichotomy the Son belongs to the first group, that of Creator, whereas in Eunomius' division, the Son does not belong to the first section which is reserved for the Unbegotten Father alone, and not for the Son who is begotten.

Eunomius does not make the Son equal with other created beings. The Son alone is the direct creation of the Father. All other things including even the Spirit, are the creations of the Son. The Son, though created, is still unique in that he alone is directly created by the Father, with the fullness of God's creative power, so that the Son came in turn, while remaining created, to be a creator also, in fact the creator of all things.

Eunomius' epistemology assumes that starting from the name, it is possible to get to the *ennoia* or intuition or mental conception of the reality signified by the name. The Father's *ousia* or essence is not unknowable. It is knowable precisely by the name *agennētos*. But the significance of the name is not in the utterance or pronounciation (*prophora*) but in the *ennoia* (the internal mental conception).

Gregory takes the opposite view in his epistemology. The names and words applied to God are human creations to denote an experience of the operation of God and not to point to the *ousia* of God.

It is clear that Eunomius has a view of the human mind characteristic of much contemporary thinking, which seems to assume that what is not knowable and statable for man is not reality at all. Gregory on the other hand has a sober awareness of the limits of the human mind, and a modest estimate of its apprehensions of reality. The Eunomian view of the human mind as potentially capable of knowing everything including the *ousia* of God can be a temptation for theologians today. The Gregorian view of the limits of the capabilities of the human mind seems at first defeatist and lazy. But as we shall see in our discussion on *epinoia* later on, Gregory has a higher and yet saner view of the human mind than that of Eunomius, whose view is self-contradictory in that it devalues the capacity of the human mind in relation to *epinoia*—especially its capacity to distinguish between the true *ennoia* and the false one. Eunomius thinks that the intuitions as well as the names given to them are created by God. Gregory says that the human mind is God's creation, but the creations of the mind may be right or wrong, good or evil.

Eunomius' basic argument in his *Apology for an Apology* has a second central element. Not only does he distinguish between the two *ousiai* of the Father and of the Son, which have nothing in common with each other, the one being uncaused and the other caused. He goes on to distinguish between the operations (*energeiai*) of the Father and the Son, which come from their different *ousiai*. Th. Dams has ably reconstructed the whole structure of Eunomius' thought. No *ousia* subsists or exists without its proper *energeia*. If the *ousia* are different, their operations are bound to be different, according to the differences of the *ousiai* from which they proceed.

Eunomius' system is quite neat. The *energeia* of the Father is the Son, and the Son alone. He alone is begotten, made, or created (it makes no difference for Eunomius which of these terms one uses) by the Father Himself. Everyone and everything else (excluding thus the Father and the Son, but including the Spirit) comes from the operations of the Son.

The Holy Spirit is the principal product of the *energeia* of the Son. So the gradation in rank is clear. The Father's *ousia* produces the Son and the Son alone. Thus the Son is superior to everything else except the Father, but in relation to the Father, he is inferior since he is *gennētōs*. Now the Spirit is inferior to the Son, for the Son is the product of a superior being, namely the Father. Proportionately as the Father is superior to the Son, the product of the Father's *energeia*, the Son, is superior to the product of the *energeia* of him who is inferior to the Father. The hierarchy is thus clear. Father uppermost, next in rank the unique Son, and third in rank the Spirit, the first of the Son's creations.

Eunomius here encounters a problem. What is the relation between the *ousia* of the Father and his *energeia?* The *ousia* of the *energeia* is not the same as the *ousia* of the *ousia* itself. The *nature* of the *energeia* is to be seen, not in the *ousia* from which the *energeia* comes, but rather in the *ergon* or work that results from the *energeia.* The Son, who is the *ergon* of the *energeia* of the Father's *ousia* does not have the same *ousia* as the Father. But the *nature* or *ousia* of the *energeia* of the Father is to be seen in the Son. Logically then, the *energeia* of the Father's *ousia*, has the same nature as the Son. That is to say, the *energeia* of the Father's *ousia* is caused, and therefore not *agennetos.* The *energeia* thus becomes a force intermediary between the Father and the Son. This intermediary *energeia* of the Father is the will of God, and the Son exists by this will. There is no rupture between the Father and the Son. The *energeia* of the Father is the connecting link, and also constitutes the *ousia* of the Son.

The Son's *energeia* then creates the Spirit, who assists the Son, mainly with the work of sanctification. The Spirit, being the creation of the Son, is thus subordinate to him. Only the Father is absolutely Lord. He alone has *monarchia.* The Spirit is subject to the Son, and the Son to the Father. The *ousia* of the Father is indivisible, simple, one, not shared with anyone else, whether Son or Spirit. It cannot be shared because it alone is *anarchos* without beginning or cause, and is in a class by itself, being the cause of everything else.

The three higher *ousiai,* of the Father, Son, and Holy Spirit do not have a common nature. But together they have made everything in the rest of creation. By the decree of the Father, the Son who is the only direct creation of the Father, with the help of the Spirit who is both the creation of the Son and his assistant, has created the universe, according to Eunomius.

The main principles of Eunomius' system, in so far as they have helped to crystallise Gregory's own thought in confronting these principles may be summarized as follows:

(1) Each *ousia* (being) is "followed" by its own *energeia* (operation).

(2) The difference between the *energeiai* can be known by observing the difference of their *erga* (works or results of their operation).

(3) The *energeiai* are proportionate to the *axioma* (worthiness or dignity) of the different *ousiai.*

(4) There is only one *ousia* who is *anarchos* (without beginning), *agennētos* (unbegotten), *aphthartos* (incorruptible), *atreptos* (unchangeable), and *akinētos* (unmoved). This *ousia* of the Father is the source and creator of all else, including the Son and the Spirit.

(5) The two basic divisions of creation are *agennētos* (unbegotten) and *gennēma* (begotten). The Son, the Spirit, and the rest of creation belong to the second class. Only the Father belongs to the first division.

(6) *Agennēsia* or unbegottenness is the *ousia* of God, as well as the name of that *ousia*. The name and the reality correspond with each other. Only He who is *agennētos* is truly God. Therefore, neither the Son nor the Spirit is God.

(7) *Agennēsia* is not a name created by the human faculty of conception (*epinoia*), for the Father is *agennētos* before the conceivers (*epinoountes*) were created. *Epinoia* is not the basis for the name *agennetos*. It is a name which corresponds with the reality, and a name created by God's providence (*Contra Eunomium*. II. 289).[13] The natures and the names

13. GNO I, p. 311/18ff. PG 45. 1008 D. See NPNF V. 277ff.

are God's creation *Contra Eunomium*. II. 403).[14] The names are suited to the realities and express their nature.

Thus one may say that the general ontological principles of Eunomius are (a) the relation between *ousia*, *energeia* and *onoma* and (b) the absolute distinction between *agēnnetos* and *gennēma*.

In terms of consistency and coherence, Eunomius' system shows considerable strength. But it denied the central affirmation of the Christian faith—that God was incarnate in Christ. For Eunomius, it is the Son who is a creation of the Father, and therefore is changeable (*treptos*) unlike the Father who is unchanging (*atreptos*) who became man. Philosophically that is more respectable than saying that the unchangeable God *became* something he was not, namely Man. The Christian doctrine that God who is unchangeable, became something he was not, namely Man, without change, is philosophically quite inconsistent.

Eunomius appears also more reasonable in his affirmation that since God cannot be caused by someone else, only the Father, who is uncaused cause of all. is alone truly God. The Christian doctrine that the *ousia* of the Godhead is in three persons—Father, Son, and Holy Ghost—is philosophically much more difficult.

Eunomius' theory that names which sound different (e.g. in different languages) do not have their significance in their sound, but in the intuition of reality to which they correspond, seems also philosophically reasonable. For example the name *agennētos* as applied to God is only a sound or collection of Greek letters. It is not the sound or the letters that correspond to the reality of God, but rather it is the intuition (*ennoia*) in our consciousness that corresponds to the nature of God. That *ennoia* or innate idea is the creation of God's providence—not a creation of man's consciousness, argues Eunomius. Philosophically, this is respectable doctrine.

It is in Gregory's criticism of this third point that his own epistemology becomes clear. Gregory denies that man has an innate idea of the *ousia* of God, and insists that he can only apprehend the *energia* or operation of God. Eunomius contends that it is only the *ousia* which is *agennētos*. The *energeia* is caused by the *ousia* and is not therefore *agennētos*. So, it cannot be the *energeia* that we understand by the term *agennētos*. It is this *energeia* which we call Father (*Pater*) or the Creator (*demiourgos*) not the *ousia*. The Father

14. GNO I, p. 344/7ff. PG 45. 1045ff. NPNF V. 290 B ff.

as *energeia* can be comprehended through its *ergon,* i.e. the Son, according to Eunomius. In this *ergon* can be found the whole creation *in principio.* Thus the creation is in the Son, and it is by the Son that all things come into existence. Becoming, or coming, to be, has thus its existence in the *energeia* of the Father, which results in the Son, who is capable of all becoming. The Son, who is *gennēma* is the source of all *gennēsis.*

The human *epinoia* can conceive the various becomings, but the *ousia* that is beyond all becoming, is understood by the *ennoia* planted in man by God, according to Eunomius. By this innate idea we can understand the basic, fundamental, primary distinction between *agennētos* and *gennēma,* and to know this distinction is to know God's *ousia.* But the distinction is not the result of reflection, but an infusion from God. (*para too theou tei phuses to toiouton katesparto*)[15].

Gregory would argue that the word *agennētos* is also the creation of the human mind (*epinoia*), arising out of human reflection. It says only what God is not, but does not explain his *ousia,* says Gregory. The word is not "innate", nor does its *ennoia* correspond with the *ousia* of God. Gregory rejects Eunomius' phenomenology of innate ideas, and proceeds to develop his own epistemology based on the principle of *analogia* or proportionality.

Th. Dams' summary of Eunomius' system recalls the essentially philosophical nature of his thought:

> "It is clear that such a theology is dominated by theses about God and the world which are not derived from Revelation. With Eunomius, there is no question of a *fides quaerens intellectum* (faith seeking understanding) of the divine mysteries. He attempts to take into account, as far as he can, the authentically Christian mysteries, but always in reliance on presupposed philosophical principles (the aspect of tradition which he seems to accept, that it is God who gives us the names of things, does not come from faith, but is a demand of philosophy)."[16]

15. *Contra Eunomium.* II. 548. GNO I, p. 386/26ff. PG 45. 1093 D. NPNF V. 305.

16. Dams, *op. cit.,* p. 129. See also p. 131:
". . . l'intention lui manque completement de contróler sa theologie par l'Ecriture. Elle est d'avance basee sur des principes determinés et en passant on remarque que ce n'est pas en contradiction avec l'Ecriture."

It is not necessary here to dwell on the Arianism of Eunomius. He uses the arsenal of scriptural texts so characteristic of Arianism, to prove the absolute transcendence of the Father alone and the inferiority of the Son to the Father. The point at which Eunomius goes beyond Arius is worth noting. For Arius God's absolute transcendence implies his unknowability. God is, for Arius, *agnōstos* (unknowable), *kai akatalēptos* (incomprehensible), *annenoētos* (unintuitable)[17] according to Philostorgius; Eunomius insists that God's *ousia* can be intuited. *Agennēsia* thus becomes for Eunomius both an ontological and an epistemological principle.

Implied in making *agennēsia* an epistemological principle, there lies a theory of perception and language peculiar to Eunomius. The theory of perception is that innate ideas correspond to reality, and the theory of language states that names and natures coincide, that names which refer to realities are created by God, along with the nature of these realities. This is certainly going beyond Plato and the "Neoplatonists". Plato would certainly not agree with Eunomius' understanding of perception and innate ideas without substantial qualifications. It may be the case as Dams suggests, that Eunomius assimilated the Stoic notion of *ennoia phusikē* (intuition of natures) to Plato's theory of ideas. The Stoics insisted only that the nature of reality can be intuited by the human mind. Eunomius qualifies this view to assert that such intuitions are created by God and are adequate for the knowledge of reality, since the names are co-created with the natures.

It is worth noting how the system of Eunomius agrees at so many points with that of Plotinus, as we will show in a moment: it is, however, at variance from the latter at two important points. First, for Plotinus the One is beyond all knowledge, as it was for Arius as well as for the Cappadocians. But Eunomius is convinced that the human mind can intuit the essence of Godhead. Second, for Plotinus there is no basic ontological discontinuity between the One and the many, since the latter emanates from the former. For Eunomius there is a basic gap between the one *agennētos* and the many *gennēmata*. He thus rejects the Plotinian theories, the unknowability of God (epistemological) and of the continuity between the One and the many (ontological).

---

17. PG 65. 460ff.
18. Dams, *op. cit.*, pp. 135ff.

This discontinuity between the One and the many is a primary philosophical problem in Eunomius. For him there is a basic ontological dissimilarity between the *ousia* of the *agennētos* and even its own *energeia*, since the first is uncaused and the second is caused. The first is One, simple, and unmoved, while the second is multiple, moving, and changeable. The *energeia* is the basis for multiplicity and change. But does this *energeia* exist apart from its *ousia*? How can there be this discontinuity between the two if one is dependent on the other? This problem will come up again when we discuss Gregory's own view of *ousia* and *energeia.*

Here it remains necessary to point out some fundamental similarities between the doctrines of Eunomius and those of "Neoplatonists" like Numenius and Plotinus. Numenius of Apamea is particularly interesting, since his Trinity bears considerable similarity to that of Eunomius. Our information about Numenius comes mainly from Eusebius' *Praeparatio Evangelica*[19] and other writings which quote him. Plotinus himself, while still alive, had to answer to charges of plagiarizing Numenius, as Porphyry tells us, and Amelius, the disciple of Plotinus, had to write a book defending his master and friend against the false charge.[20]

Numenius has three first principles—the One, the Intellect and the All-soul. Plotinus has also three. So has Proclus, who comes later than Gregory in time. The Trinity of these neoplatonists is not exactly identical. For Plotinus, as for Numenius, the First Principle is the Good, One (*hen* for Plotinus, *monas* for Numenius), simple, transcending all being and all works. He alone is *autoon* (himself being being). Since we have more complete documentation for Plotinus than for Numenius, let us compare Eunomius with Plotinus.

The first tractate of Plotinus' fifth Ennead deals with the three initial hypostases.[21]

> "Because thus it is necessary to reckon, that above Being, there is the One, which we have sought to demonstrate in as

19. *op. cit.*, XI. 18. 20. See E.A. Leemans, *Studie over den wijsgeer Numenius van Apamea met uitgave der fragmenten* (Mémoires de l'Académie Royale de Belgique, Classe des lettres, xxxvii. 2), Brussels 1937, which gives all the available fragments.

20. *Vita Plotinii.* 17. Bréhier's Edition Tom. 1, pp. 17-18.

21. The title is *Peri tōn triōn archikōn hupostaseōn.*

> far as such matters are demonstrable; next that which is both Being and Mind *epheyēs to on kai nous),* and third the Nature of the Soul *(tritē de hē tēs psuchēs phusis."*[22]

This is very close to Eunomius' teaching as cited by Gregory in *Contra Eunomium.* I. 151.

> "The whole summary of our teachings is concerning the Highest and Most Sovereign Being, (*ek te tēs anōtatō kai kuriōtatēs ousias*) and that being which exists on account of it, but after it, though senior in being to all other beings *(ek tēs di' ekeinēn men ousēs met' ekeinēn de pantōn tōn allōn proteuousēs)* and a third, not to be put in the same rank as the two others, but subordinate to them, to the first because of the difference in cause, to the second because of the difference in the *energeia* which produced it, taking into account, of course, the *energeiai* which are consequent to each *ousia*, and the names which are attached to these."[23]

We see here a close parallel between the two teachings, though the way of putting it is extremely personal for Plotinus as for Eunomius. Eunomius is no mean plagiarist. He has assimilated the teaching of Numenius and Plotinus. The three higher hypostases of Eunomius correspond exceedingly well with those of the Neoplatonists. Neither of these correspond to the homoousian Trinity of the Nicene Party as taught by the Cappadocians.

It is then reasonable to suggest that Gregory's intellectual effort is directed against Christian doctrine being too uncritically adapted to Neoplatonist doctrines of the pagans of Alexandria, or to the Jewish Platonism of a Philo, also of Alexandria. It seems to the present writer that Gregory was fighting against the wrong kind of indigenization of Christian theology within the forms of contemporary culture—wrong in that it took the axioms of contemporary philosophy rather than the affirmations of the faith regarding the Incarnation and the Trinity, as basic criteria. The kind of indigenization that Eunomius attempted was characterized by a low regard for the Tradition of the Church and an overhigh and uncritical respect for

22. *Ennead* V. 1. 10. Bréhier's Ed. Tom. V, p. 28, line 1ff.
23. GNO I, p. 71/28 to p. 72/10. PG 45. 297. NPNF V. 50a.

contemporary philosophical trends. Such an approach is still a temptation for many modern theologians, though quite often today, theologians are not as philosophically astute and as logically consistent as the "technologue" Eunomius.

1. Eunomius—Theurgic Neoplatonist?

It is not our intention here to make a comparative study of Eunomius and the Neoplatonists, though such a study would be eminently valuable. Enough has been said to show that Eunomius was more of a Neoplatonist than Gregory of Nyssa. Eunomius was not a simple adapter of the thought of Plotinus (205-270 A.D.). He belongs to a later stage in the development of Neoplatonism. Porphyry (233-303 A.D.) and Amelius (246-270 A.D.), both disciples of Plotinus, and Longinus of Emesa (213-273 A.D.), teacher of Porphyry and critic of Plotinus, had made the thought of Plotinus known in Rome and Athens.[24]

The religious history of the third and fourth century Eastern mediterranean societies remains yet to be comprehensively studied. But it is conceivable that what took place at Alexandria was not an isolated phenomenon. There was a religious development outside of the Judaeo-Christian tradition within the Eastern mediterranean world, of which Ammonius Saccas, Plotinus, Porphyry, Iamblichus and Proclus form but one line of development. There were other lines of which we know definitely, take for example the Asiatic or Stoic Line of Zeno-Cleanthes-Chrysippus-Poseidonius-Cicero-Epictetus-Marcus Aurelius. Here again Asian spirituality is naturalized on Graeco-Roman soil, and by the third century Stoicism becomes the religion of the establishment—the religion of good citizenship—a far cry from the earlier Stoicism which was primarily a way of protest, endurance and martyrdom.

By the fourth century Stoicism hardly survives as a distinct religious system; what we have is a much more free pattern of religious thought in which elements from the various schools are combined, but interpreted according to the personal style of individual teachers

24. Longinus wrote a review of Plotinus' work *On the Ideas*, in reply to an account of this work by Porphyry. Longinus disagreed with Plotinus basically, and obviously developed a rival system. His *Letter to Amelius* is also a criticism of Plotinus, in answer to Amelius' work *On Plotinus' Philosophic Method*. *Vita Plotinii*. 20. Bréhier, vol. I, p. 23.

and writers. This is what we see in the case of Libanius, Himerios, Themistios or Julian the Apostate, all four contemporaries of the Cappadocian Fathers.

But the system of Ammonius Saccas has its own distinct continuity, again with Asians as leading teachers. Iamblichus of Apamea (ca. 250-ca. 325 A.D.) lived well into the first quarter of the fourth century. However shallow or mistaken he may have been in his commentaries on Plato and Aristotle, it seems quite possible that he added an element to Plotinian philosophy which was totally absent there—the concept of "mystery" as worship of a secret community of initiates. This is understandable in the light of what we know about the vogue of mystery cults in that time and clime. There was a Plotinian School at Athens as well, which later Proclus (410-485 A.D.) was to lead.

Did Eunomius belong to one of these Plotinian Schools, especially one of the Iamblichus type, which incorporated into Plotinus' philosophy some form of a mystery cult? The question is interesting, but no positive answer seems possible with the information now available. In any case it is clear that the epistemology of Proclus, head of the Athenian school of Neoplatonism, is very close to that of Eunomius. Proclus had a high view of words and believed that names had been created along with things, and names are essential for constructing a logical structure parallel to the ontological structure "Words are the imitations (*mimēseis*) or images (*ikones*) of things (*pragmata*) and are correlated with what they imitate."[25]

Here we follow Mme Abramowski[26] to a large extent. She comments on le Bachelet,[27] who wrote in the early decades of our century questioning the earlier tradition of regarding Eunomius only as a Sophist dialectician who uses the *technologia* of the Second Sophistique. But le Bachelet did not take into account the relation between the two cardinal points of Eunomius system—i.e. *agennēsia*

25. *Commentary on Cratylus*. 68.

26. Reallexikon. *loc. cit.*

27. le Bachelet wrote the article on *Eunomius* in *Dictionnaire de Théologie Catholique*, vol. V. 2. cols 1501-14.

28. J. Danielou also refers to this dissertation.

See his *Eunome l'Arien et l'exégèse neo-platonicienne de Cratyle* in *Revue des Etudes Grecques* Tome LXIX (1956), pp. 412-432. The present writer has been able to secure a copy of this typescript and acknowledges his indebtedness to Dams' work.

or unbegottenness as the *ousia* (essence) of Godhead, and the assertion that names and words are of divine origin and not to be tampered with.

Mme Abramowski asserts that Eunomius' epistemology is not Aristotelian. She refers to the work of Th. Dams[28] on the Eunomian Controversy in which he has reconstructed the architechtonic and consistent structure of Eunomius' thought. Danielou has also used this material to come to an interesting conclusion.

The idea that names (especially of Gods) in each language, whether Greek or Barbarian, are of divine origin and not to be tampered with can be found in the neoplatonic commentaries on Plato's *Cratylus*, as is clear from Proclus who lived and wrote a little later (410-485 A.D.) than Gregory and Eunomius. This idea of Eunomius has its origin, according to Danielou,[29] in the thought of Iamblichus (ca. 250-ca. 325 A.D.), who is often credited with introducting the Theurgic or liturgic element into Neoplatonic mysticism. Danielou thinks that Eunomius frequented the school of Iamblichus and his disciples and picked up this central idea of his epistemology there.[30] He contends that Eunomius is an original thinker who altered the Neoplatonic system by the use of Aristotelian categories:

> "The characteristic of Eunomius is thus to combine with a mystical platonism influenced by theurgy, a philosophical technique which is chiefly Aristotelian."[31]

---

29. Danielou, *Eunome l'Arien*, *op. cit.*, p. 427.

30. Danielou is perhaps more categorical than is warranted by the facts : *"Le système d'Eunome est en fait un système néo-platonicien, une explication de la génèse du multiple á partir de l'Un."* op. cit., p. 428.

31. *op. cit.*, p. 429. By *theurgy* is meant a supra-rational non-sacrificial cultus of God, a form of worship which seems absent in Plotinus, whose "mysticism" was an intellectual one combined with a strict personal discipline, but without any elements of worship. Plotinus had of course "mystical" experience of the soul's union with God, but no corporate or ritual form of worship was prescribed by him. Porphyry sought to combine theurgic worship with Plotinian mysticism. The source of this theurgic element is no doubt the *Chaldean Oracles* believed to have been written by one Julianus during the reign of Marcus Aurelius (121-180 A.D.). Porphyry wrote a treatise *On the Philosophy of the Oracles (Peri Tēs ek Logiōn Philosophias)*, of which the Chaldean Oracles seemed to have formed the nucleus. (See H. Lewy, *Chaldean Oracles and Theurgy*, Recherches d'Archéologie, de Philologie et Histoire, XIII. Le Caire 1956). The purpose of the theurgical rites was to purify the

"On the other hand," continues Danielou,

> "he is a Sophist. He is of course contemporary of Libanius and Himerios . . . but he also presumes to be a mystagogue, who speaks in very solemn tones . . . With Eunomius. Arianism takes on a quasiesoteric character. Eunomius is the hierophant of a gnosis, of a secret doctrine, which is true piety. He considers Basil as a non-initiate (*amuētos—Contra Eunomium*. I. 74. 44.20). He is thus the contemporary of a neo-platonism which is not a simply a metaphysical system, but also a theurgy. He stands thus as an intermediary between Christianity and this kind of neo-platonism."

But what is meant by "theurgy" in such a context? *Theourgia* is a word used by Julian[32] to denote a divine work, or as in Porphyry[33] a sacramental rite or *musterion*. Iamblichus[34] also uses it in this sense. It came to mean, in later fathers, sanctification and deification through the sacramental mysteries.

Here is a problem that Danielou does not take into account. The point at issue epistemologically between Eunomius and Gregory is whether "the mystery of piety" is contained in the sacramental symbols or in precise dogma. Eunomius takes the view that precise dogma is the true expression of faith and not the sacramental mysteries or terms used in worship. His words are clear:

> "But we (Eunomius), persuaded by holy and blessed men, affirm that the 'mystery of Godliness' is neither in the dignity of names, nor in the specificity of practices and sacramental mysteries, but rather in the exactness of teachings."[35]

---

pneumatikē psuchē and to lead it to perfection. Augustine seems to have been significantly influenced by Porphyry's work on Theurgy in a Latin translation. (See John J. O'Meara, *Porphyry's Philosophy from Oracles in Augustine,* Paris 1959). The theurgical school of Neoplatonism antedates Iamblichus.

32. Julian the Apostate, *Oratio* 7. 219a (ed. F.C. Hertlein, Leipzig 1875-76). See also Dionysius the Areopagite, *Celestial Hierarchy* 7. 2 (PG 3. 208 C, D).

33. Porphyry cited by Augustine in *Civit. Dei*. X. 9.

34. *de Mysteriis*. 9. 6. (ed. G. Parthey, Berlin 1857).

35. *Hēmeis de peithomenoi tois hagiois kai makariois andrasin oute tei semnotēti tōn onomatōn oute ethōn kai mustikōn sumbolōn idiotēti kurusthai phamen to tēs eusebeias musterion tei de tōn dogmatōn akribeia. Contra Eunomium*. Lib. III. Tom. IX. 54. GNO Vol. II, p. 284/20 ff. PG 45: 877 D. NPNF, V : 238 A.

Gregory does not deny the importance of doctrines, but he would deny the statement that they constitute the essence of the faith. The essence of the faith for Gregory is to be born of water and the spirit and to eat the flesh and drink the blood of Christ.[36] He insists that heretics and unbelievers have sometimes "hit" the right doctrine, though most of the time they are in error. But even after succeeding in "hitting" the right doctrine, they remain estranged from the faith,[37] so long as they remain *exō tou mustēriou* (outside the mystery). In denying the meaning of the baptismal formula and the validity of the sacramental symbols, Eunomius is denying the very Christian Faith. Gregory's language becomes very strong at this point :

> "Since then he despises the revered names through which the power of a more Godly birth imparts grace to those who approach in faith, and since he disregards the communion of the sacramental mysteries and Church customs in which Christianity has its strength...how is it that you do not see (in him) the persecutor of the faith persuading you to agree with him in betraying Christianity"?[38]

The difficulty then, in accepting Danielou's hypothesis that Eunomius' neoplatonism was a theurgic one, remains quite considerable. For Eunomius "dogmatic precision" is more important than any sacramental mystery.

2. **Epinoia**

The fundamental epistemological point of Eunomius relates to the concept of *epinoia* which is sometimes translated as "conception" and can in fact mean the human capacity to conceive things which exist or do not exist. The locus classicus of Eunomius' concept of *epinoia* is cited by Gregory in full.

> "Here is a logical master-piece of the dialectical technique, such as none among those uninitiated into its hidden wisdom

36. See GNO II. 285/5 ff.
37. GNO II. 285/12-17.
38. GNO II. 285.

> will dare to encounter! But in so solemnly explaining to us the varieties of 'conceptions' (*epinoias*) he makes quite a funny story of the very concept conception. Here are his words: 'Of those things called conceptions, some have their existence only in the utterance, as for example those that signify nothing; others exist only in the thought of the particular person; and of those some are (created) by enlargement as for example the Colossi (giants), others by diminution, as for example the pygmies; others by addition, e.g. multi-headed animals, and yet others by combination as for example Centaurs."[39]

Gregory laughs at Eunomius for picking up some of the worst examples of the creations of the human imagination and trying to understand the *epinoia* or the human capacity for thought in such ridiculous terms.[40] To this Gregory opposes his own very lofty, perhaps too lofty, understanding of the capacities of the human mind :

> "Whence cometh then the higher disciplines, whence geometry? When the theoretical teaching of philosophy, and logic and physics, the invention of machines, the marvellous devices of chronometry through the brass-dial and the water-clock?[41] So also whence the philosophy of being, the

39. *Contra Eunomium* II. 179. GNO I, p. 276/18 ff.
Migne's Bk XII. PG 45. 969. NPNF. V: 267 B.

40. *Ibid.*

41. The water-clock was one of the wonders of Alexandria. Vitruvius (IX, VII. 4 and 5) describes the technique of this clock: "For the water-inlet, he (Ctesibius, the son of a barber, 285-222 B.C.) used a hole bored in a piece of gold or in a gem, finding that these neither wear out nor get blocked. This secured an even flow. The water as it rose floated an inverted bowl, technically known as the cork or drum. This connected with a bar and a revolving drum. Both bar and drum had teeth at regular intervals fitting into one another. By this means the rectilinear motion of the rising cork was transformed into a series of small measured circular movements. By an elaboration of this device through a number of rods and cogs he caused a variety of movements. The little (human) figure which pointed to the hour moved. The cylinder of the clock revolved. Stones or eggs were dropped. Trumpets were sounded, and there were other incidental effects."
quoted by Benjamin Farrington, *Greek Science*, Penguin, vol. II, 1949, pp. 61-62.

> knowledge of ideas, in one phrase, all the soul's dealings with higher and lower realities? What about agriculture, navigation and the other different activities of human life? How does the sea become a highway for man? How do the beings of the air come to serve those that inhabit the land? How are wild beasts domesticated, those that infuse terror subdued? How is it that animals stronger than men are brought under human control? Have not all these accrued to human life by virtue of the human *epinoia*? According to us, then, *epinoiā* is the method of discovering things unknown, through the following up of the first notion regarding the object of study, searching out the adjoining steps to advance to further discoveries.[42] For, having conceived something regarding the object under investigation and harmonizing with this first idea other notions discovered, we gradually find our way to the limits of the studied objects."[43]

The human *epinoia* is thus the instrument by which we seek out, through the process of following up (*akolouthia*) our notions, and attain to inventions as well as to theoretical knowledge. All human knowledge is achieved through the *epinoia* which Eunomius so uncomprehendingly despised and ridiculed.

According to Gregory, however, "...all things whatever time has discovered as useful and beneficial for the life of men have been discovered in no other way than through the *epinoia*."[44] It is the most precious of all the faculties given by the Creator to man in the divine Providence. Just in case someone accuses Gregory of being too hellenistic at this point, he quotes Scripture to prove that his idea is fully biblical, and cites Job 38:36[45] which says that it is God who has set man over technique and given to woman the knowledge of weaving and embroidering. Gregory did not have the Hatch and Redpath concordance to the Septuagint at his disposal. If he had,

42. This is Gregory's epistemological principle and he defines epinoia here : *esti gar kata ge ton emon logon hē epinoia ephodos heuretikē tōn agnooumenon dia tōn prosechōn te kai akolouthōn tei protei peri to spoudazomenon noēsei to ephexēs exeuriskoumena.*

43. *Contra Eunomium* II. 181. 182. GNO I, p. 277/7 ff.
Migne's Bk XII. PG 45. 969. NPNF V, p. 268 A.

44. *Ibid.*

45. LXX.

he could have seen that the word *epinoia* occurs in the LXX in a moral sense,[46] and he would have strengthened his argument that the *epinoia* is as capable of false as of true conceptions. But that is precisely Gregory's argument here and his epistemological criterion. Gregory makes the following points on *epinoia:*

(a) It is the faculty of the mind that goes beyond sense impressions (*aisthēsis*) to relate them to each other in the mind and thus forms concepts.
(b) All human advance and progress is dependent on the proper exercise of this faculty and as such it is the greatest gift of God to Man.
(c) This faculty, however, is not autonomous, but given by God, and functions properly only under God's guidance.
(d) This faculty can err; but this is essentially the character of all human faculties, that they can be used for opposite purposes, i.e. for the truth or for falsehood, for good or for evil. (The doctor can heal or kill; the captain of a ship can steer clear of the rocks or destroy the ship by running it into the rocks—and so on for the painter, wrestling master, etc.) The faculty is implanted in man for good, but man has the moral freedom to use it for the opposite. Its capacity for evil and falsehood is only the obverse of its capacity for the good and the true. Instances in which it is used for evil or falsehood cannot be used to disprove its capacity for what is good and true.

Gregory summarizes his argument against Eunomius thus:

> "The fact that the mind (*nous*) implanted in us by the Provider is perfectly capable of inventing non-existent beings does not imply that it is endowed with no power or faculty (*dunamis*) capable of providing discoveries that are beneficial to the soul. As the impulsive and choosing faculty (*dunamis*) of our soul (*psuchē*) has been built into it for striving after the good and the beautiful, in accordance with the inner leadings, though one can also use the same impulse for that which one ought not,

46. II. Maccabees 12 : 45; IV Macc. 17 : 2. Wisd. of Solomon 6 : 16, 9 : 14, 14 : 12, 15 : 4, Wisdom of Sirach (Ecclesiastius) 40 : 2 and Jeremiah 20 : 10.

> no one can cite the propensity of choice towards that which is evil as proof for its having no propensity for the good; thus the movement of the epinoia towards that which is vain and unprofitable is not evidence (*kategoria*) for its incapacity for that which is profitable, but becomes sure proof (*apodeixis*) for its not being incapable for that which is spiritually beneficial and necessary."[47]

The word *agennēsia* or unbegottenness which Eunomius regards as a sacred term of divine origin to denote the very *ousia* or essential being of God, is now clearly stated by Gregory to be a human creation to express one of the conclusions to which the human *epinoia* comes in this process of moving forward to the good and the true. In our search for the first cause of all existent beings, that is, for God, it becomes clear to us that He who is the cause of everything else cannot Himself be caused by something else. Our *epinoia* creates the term *agennētos* or unbegotten as an apellation for Him who is without beginning. But the name, says Gregory, does not describe him; it does not say what he is, but only denotes what He is not.[48]

This is both the possibility and the limit of the *epinoia.* It can proceed, through the process of *akolouthia* which we shall elucidate later on, through all the existent beings (*ta onta*) to the anterior cause which thought conceives to be necessary.[49]

This is the limitation of all created knowledge; but the limit does not make it useless or unworthy. On the contrary the human *epinoia* is the instrument through which all science, technology and art is achieved, says Gregory. And that is no achievement to be sniffed at, either for Gregory or for us. Its legitimate function is to

---

47. *Contra Eunomium* II. 191. GNO I. 280/6ff. PG 45. 973. NPNF V. 269 A.

48. *Ton de houtōs onta agennēton kai anarchon ōnomasamen, ou ti estin, alla ti ouk esti dia tou onomatos endeiknumenoi. Contra Eunomium* II. 192. GNO I. 280. 27ff. PG 45. 973—NPNF V. p. 269 A.

49. This is the "God of the gaps" of Bonhoeffer, the *Deus ex machina* of human thought. But human reason cannot know this God, except see that He must be. The "God of the gaps", however, is not a *creation* of the *epinoia*, but its *discovery*. The epinoia cannot tell us who God is, but only show us that while his facticity or is-ness is recognizable his essence or what-ness remains beyond the capacity of the conception.

follow up (*akolouthein*) the creation in its orderliness (*taxis*) and mutual relationship, and to show us that the ground of all existence in creation lies "beyond" the creation and can be affirmed though not described. The *epinoia* can discover the orderliness of creation as well as its nature as contingent upon the Creator's will and energy, but it also reveals the gap or *diastema* between the Creator and the creation. Bonhoeffer would not have been justified, had he read Gregory, in calling Gregory's God the "God of the gaps", for God is not in the gap, but the gap is simply an epistemological-ontological reality that confronts man in his attempt to know God.

A clear understanding of this notion of the gap or interval in Gregory's thought is thus essential to understanding the formal structure of Gregory's ontology and epistemology.[50]

50. See chapter on God and His creation I. (Chapter VIII).

CHAPTER III

AKOLOUTHIA—FOLLOWING THE LOGOS
The Epistemological Method which Combines Moral, Logical and Spiritual Life in One Single Quest for Meaning

The term *akolouthia* which could be translated consistency, consequence, coherence, following up or custom, occupies a central place in Gregory's epistemology. The word and its derivatives are used frequently in practically all his works.[1] Significant seems the fact that occurrences of these words are more frequent in the philosophical writings.[2] The word does not appear to have had the same centrality in the earlier philosophical tradition, whether Platonic or Aristotelian.

Philo Mechanicus (3rd to 2nd century B.C.) had already used the word in the sense of logical consequence[3] and in the third century B.C. Chrysippus the Stoic (281-208 B.C.) uses it in the same sense,[4] and Gregory must have taken the term from Stoic sources.

---

1. In the *Contra Eunomium* alone the noun is used 124 times, and the adjective or adverb 84 times. I have not tried to count the number of times the verb is used.

2. For example *In Inscriptio Psalmorum* uses *akolouthia* 32 times and the adjective or adverb 56 times, *In Canticum Canticorum* 19 and 37 times respectively.

3. *Belopoeica, Excerpte aus Philons Mechanik,* ed. H. Diels, E. Schramm, *Abhandl. Berlini. Akadem.* 1919, No. 12. cited in Liddell & Scott, Greek-English Lexicon, under *akolouthia.*

4. H. Von Arnim, etc. *Stoicorum Veterum Fragmenta,* Leipzig 1903, vol. 2. 68 and elsewhere, cited by Liddell & Scott, *op. cit.*

One finds the words *akoloutheō akolouthia* and *akolouthon* of frequent occurrence in treatises of Stoic logic. See Diogenes Laertius, *Vitae Philosophorum* (Ed. C.G. Cobet, Paris 1878) VII. 71, 74, 81 etc. (Eng. Tr. in Diogenes Laertius *Lives of Eminent Philosophers* Trans. R.D. Hicks, 2 vols, Loeb, 1925) and Sextus Empiricus, *Against the Mathematicians* (Eng. Tr. R.G. Bury, Loeb, 2 vols, 1917+1933). VIII. 111, 112 etc.

Cardinal Danielou has pointed to the central meaning of the term for Gregory as a necessary causal consequence which explains something.[5] The Stoics had used it not only in a logical sense, but also with a moral meaning. *Akolouthōs tei phusei zēn* meant for the Stoics to live according to nature, i.e. in accordance with the Logos that indwells the cosmos and all things in it.[6] Philo (Alexandrinus) had also a notion of a fixed order according to which everything proceeds, though he does not always use the term *akolouthia* for that order. It is sometimes *taxis*[7] (a cosmic order, which in Aristotle fixes the length of life of a man), sometimes *tuchē* (which is fortune as a form of God's providence.)[8] But *taxis* or order can also mean for Philo rank and sequence. In fact he defines *taxis* in terms of *akolouthia* (sequence) and *heirmos* (linkedness).[9] There is a certain sequence in the very creation, which cannot be reversed, according to Philo.[10]

For Gregory, it becomes of the utmost importance to understand things in their proper sequential order, and then to state what is understood in the right sequential order. Good discourse is what manifests good *akolouthia.* Sometimes he uses the word in a much more ordinary sense, like for example meaning the course of life[11] or the course of nature,[12] at other times it is in Danielou's happy phrase, an *enchainement,* a chain, often of evil,[13] or of sin.

In the controversy with Eunomius, however, the idea takes on a fundamental significance. Eunomius argued that the Son and the Spirit came later in the sequence of revelation and therefore that they are inferior to the Father who comes first in the sequence of revelation. This argument Gregory refutes very easily.[14] One cannot say two things at the same time, and if one says one thing

5. J. Danielou, *L'être et le temps chez Grégoire de Nysse,* p. 19.

6. See Danielou, *Platonisme et Théologie Mystique,* p. 302 note; also Völker, *Gregor von Nyssa als Mystiker.*

7. See Aristotle, *De Generatione et Corruptione* II. 10, 336 b. 10-13.

8. H.A. Wolfson, *Philo* I, p. 330.

9. *De opificio Mundi*, 7. 28.

10. See Wolfson, *Philo* I, pp. 215-216.

11. *Akolouthia tou biou, Contra Eunomium* I. GNO I. 34.10. PG 45. 260 B.

12. *Akolouthia tes phuseōs, In Canticum Canticorum*: Hom. 13. GNO VI. 389 line 3. PG 44. 1053 B.

13. *Tōn kakōn akolouthia, Contra Eunomium* I. 89. GNO I. 53. 5-6. PG 45. 277.

after the other, it does not necessarily follow that the latter is inferior to the former.

Gregory's point is that *akolouthia* or order and sequence belongs only to the created order and not to the Creator. In the created order it is possible to find order and sequence and know things in terms of their mutual relationship. But in the uncreated there is no sequence or rank, and our method of knowing things, which Gregory sees as discovering the *taxis* and *akolouthia* of things, cannot go beyond the created order.[16]

But this is by no means a disparaging of *akolouthia* for only through discovering relationship between notions we can approve or reject statements about God such as those made by Eunomius. Eunomius' contention, for example, is that the Father alone, as ungenerate (*agennētos*) has being in himself (*auto-ousios*) and is good in himself (*autagathos*). The Son and the Spirit have their being from the Father, and are inferior to him in both being and goodness, according to Eunomius.

What is the *akolouthia* of the argument of Eunomius? "Let us mark the consequence" (*skopēson tēn akolouthian*). If the Son has goodness only by participation and not in himself and is "less perfect", then it follows, according to Gregory, that the nature of the Son is lacking in some degree in goodness, has to change into a greater degree of goodness for himself in order to be perfect, is therefore not God, since in need of becoming; and therefore, argues Gregory in a very sophistic and rather unconvincing syllogism, he cannot save the world, being in need of something himself.

But this argument has its own strength when further analysed. If the Son is less than the Father, then the Son is not infinite. If he is not infinite, then his growth into the infinite has to be also an infinite process, for it is axiomatic for Gregory that the finite can never become infinite by addition, but has to be infinitely in the process of being added to.[17]

Here Gregory uses the notion of *akolouthia* to test certain doctrines by working out their logical consequences. Whether a

14. *Contra Eunomium* I. 197ff. GNO I. 84/10ff. NPNF V, pp. 53-54.

15. See *Contra Eunomium* I. 270-272. GNO I. 105/19—106/11. PG 45. 333 B.C.

16. *Contra Eunomium* I. 288. GNO I, p. 111/17ff. PG 45. 340. NPNF V. 62 A ff.

17. *Ibid.* (note 17)

doctrine is true or not is to be tested in terms of its coherence or non-contradiction with other well-attested facts and doctrines.

Akolouthia is thus an aspect of the orderliness of the universe, as Philo Alexandrinus had already stated three centuries earlier: "The good can never be in disorder. Order (*taxis*) is sequence (*akolouthia*) and coherence (*eirmos*) of that which precedes and that which succeeds, if not also in the effects, at least in the conception (*epinoia*) of the architect."[18]

In Philo's conception, drawn from Stoic sources, *akolouthia* has to do with the design of the created order in the mind of the Creator, and is thus an ontological reality, with epistemological consequences. We are to understand the *akolouthia* or *taxis* of the universe if we are to understand anything real at all.

There can be little doubt that Gregory's *In Hexaemeron* draws heavily on Philo's *De opificio Mundi.* But Philo is more like Basil, even somewhat less attentive to philosophical coherence.[19] Gregory systematises these thoughts of Philo, which have also been dealt with by Origen,[20] into a more coherent system.

In the mind of God, being and knowledge coincide, according to Gregory :

> "In the divine nature, power (*dunamis*) is concurrent (*sundromos*) with will (*boulesis*), and the measure of the power of God is the will. But the will is wisdom (*sophia*). And the property of wisdom is not to be ignorant of how every single thing came to be. With the knowledge also springs up the power to achieve. So whatever is known also immediately has to come to be, the power to work out beings being concurrent; to be known leads to being effected, nothing intervening between known and coming into existence, but immediately, without any interval (between intention and achievement) the work becomes accomplished simultaneously with the will (*sunēmmenōs kai*

18. *Kalon gar ouden en ataxia. Taxis d'akolouthia kai heirmos esti proēgoumenōn tinōn kai epomenōn ei kai mē tois apotelesmasin, alla tais tektainomenōn epinoias.* Philo: de opificio Mundi: 28.

19. Philo does say, for example, that it would be "unphilosophic" to declare that time is more ancient than creation *(presbuteron d'apophainesthai tolman aphilosophon)—De opificio.* 26) but he is not always as careful to maintain this philosophical coherence as Gregory.

20. See *Commentary on St. John.*

> *adiastatos sunanadeiknutai tēi boulēi kai to ergon).* Thus the will is effective power, so that whenever beings are willed they also come into being; those which are conceived have the possibility and occasion for their existence in being willed. From God's side all things in creation are to be understood together—the will, the wisdom, the effective power and the essence of existing beings. (*hōs homou ta panta tou theou peri tēn ktisin noēisthai. to thelēma, tēn sophian, tēn dunamin, tēn ousian tōn ontōn.*"[21]

The fundamental epistemological principle of Gregory's thought is here made clear. God's wisdom and His will are coterminous. What He wills is what is wise, and by the very act of God's knowing and willing, everything receives its being.[22]

---

21. *In Hexaemeron.* PG 44. 69 A.

22. The sources of this conception of Gregory, that the wisdom *(sophia), thelēma* (or *boulēsis*) and power (*dynamis*) of God are the ground of all existents, can probably be traced back to the Chaldean Oracles, (GNO I: 250) and Gregory conceivably got it from Porphyry's "*Philosophy from Oracles*". The Chaldean system, as interpreted by H. Lewy in *Chaldean Oracles and Theurgy* (Recherches d'Archéologie, de Philologie et Histoire XIII, Le Caire 1956) goes as follows: "The Chaldean Hierarchy is headed by a Supreme God, whose mythical predicates are: 'Father' . . . the personality of the 'Father' remains transcendent, but his action unfolds itself through His Power. Power (*dunamis*) is the sum of noetic potencies included in the Supreme God . . . The Father himself has no direct external activity; He uses intermediaries. These intermediaries are His faculties who in their virtuality are identical with the Supreme Being, but acquire in the state of actuality a particular existence. The action of the transcendent God is thought, consequently the first entity that issues from Him is His intellect, the nous. His will (*boule*) acts in harmony with this entity, for His volition is thought, and His thought is action. Intellect, Will and Power constitute as the immediate faculties of the 'Father', the 'Paternal Monad'. . . His will and His intellect cooperate in the unfolding of His power. Every act of thought in the unfolding of the Paternal Intellect brings forth new noetic orders, described as procreations." (cited by John J. O'Meara in *Porphyry's Philosophy from Oracles in Augustine,* Etudes Augustiniennes, Paris 1959). This work of Prof. O'Meara advances the thesis that Porphyry's *Philosophy from Oracles (Peri tes ek logiōn philosophias)* was the work of the "Platonist" which Augustine mentions in Bk X of Civitate Dei as *De Regressu Animae,* which was occasion for Augustine's conversion. (See *op. cit.,* p. 145). Of course Gregory, who accepts this insight from the Chaldean Oracles, rejects the system of that work, which is similar to the Neoplatonic Trinity. For Augustine's reaction to this non-Platonic Chaldean element in Porphyry, see *De Civitate Dei* XIX: 27. For fragments from the work of Porphyry see Eusebius' *Praeparatio Evangelica.* IX. 10.

Now God's wisdom has a particular *taxis* or order. To know and to follow this *taxis* by *akolouthia* is to participate in God's wisdom which is the foundation or ground of the being of all that has existence. Here Gregory's thought seems to have been influenced by two factors at least, though both of them could have come to him by the same channel. One is the Platonic heritage, and the other the Chaldean heritage (see note 23).

In the Platonic tradition words like *nous* and *dianoia* have more than an intellectual meaning. *Dianoia* is both an idea in the mind and the intention to realize it, thus a joint act of knowing and willing.[23] Seneca, the Latin thinker, had to make a distinction between intellect and will, but the Greek language does not have exact parallels for the Latin *mens* and *voluntas.* But even in Seneca, the separation between knowing and willing is not so neat. *Velle* (to will) even in many Latin writers, means both to know what is good and to desire for it. The mind and the will are not water-tight compartments in the old traditions, and it is only in our alienated society that we have learned to develop knowledge divorced from will, calling it "objective" or "scientific". Virtue for Seneca and the Stoics as well as for Plato and the Platonists is the combination of *scientia* and *voluntas*, of knowledge and will.

Akolouthia, or "following" thus becomes both intellectual and voluntative in the same act. We follow the order of reality intellectually in order that we may follow it in terms of our life and action and therefore ontologically.

This linking ontology, epistemology and ethics, characteristic of Stoic thought as well as of Gregory's thinking, was a central notion in Christian patristic thought. Werner Jaeger refers to the ancient idea that the philosopher is a "Man interested in God", and cites Justin Martyr's Trypho the Jew meeting a Greek philosopher and assuming that because he is a philosopher, he is interested in God[24]. But this "interest in God" is far from an academic one. We are "interested" in many things because discussing them gives us pleasure. But for those men the thirst for knowledge came from the thirst for being. *Akolouthia* is both "following" intellectually and "following" as discipleship. But it is not a question of first

23. See M. Pohlenz, *Philosophie und Erlebnis*, pp. 112-113. *Die Stoa*. II. 139-140.

J.M. Rist. *Stoic Philosophy*, p. 223.

knowing and then doing. It is only as there is progress in virtue that knowledge itself becomes clearer and clearer (but not necessarily more conceptually articulable).

1. Akolouthia as Logical Consistency

The clearest example of Gregory's view of *akolouthia* as logical consistency is in his *Antirheticus against Apolinaris*. Looking at the logical consequences of Apolinaris' teaching that Christ's flesh was from heaven, Gregory argues how this leads to the conclusion that the Incarnation took place already before the creation of Adam, whereas the Church's teaching is clearly that the Logos who was in the beginning with God, in these last days for us and for our salvation became man from Mary, of the lineage of David. Apolinaris argument thus by necessity leads to logical inconsistency (*hē akolouthia tēs atopias tōn muthōn kat' anagkēn proagētai*).[25] One sees the use of the same technique in refuting the heresy of the Macedonians in *De Spiritu Sancto*[26] as well as in the controversy with Eunomius. It is safe then to say that Gregory, following up the logical consequences to see whether they lead to inconsistency (*atopia*) constitutes a basic epistemological technique to test statements of truth made by others.

The human *logismos* or rationality may be unable to comprehend the uncreated Being of God,[27] but is certainly component to shift truth from error, especially if it remains rooted in the faith of the Church.

2. Akolouthia as Ontological Reality

The capacity of the mind to follow reality has its source in the consistency and coherence of that reality. Logical coherence is possible because of ontological coherence. The ontological ground of what exists is the *logos* who proceeds from God, in accordance with a certain order or arrangement that is coherent. In his reply

---

24. Justin Martyr, *Dialogue with Trypho*.

Jaeger, *Early Christianity and Greek Paideia*, p. 31.

25. GNO III. 1, p. 150/8-9. PG 45. 1149 D. See also 148. 18ff. PG 45. 1148 D.

26. GNO III. 1, p. 95/16, 102/23, 104/31, 107/15 etc.

PG 45. 1312 A, 1320 D, 1324 B, 1328 B, etc.

27. See e.g. *Contra Eunomium* I, 367. GNO I, p, 135/13ff.

to Eunomius' Second book, Gregory elaborates this point in the context of refuting Eunomius' argument that names or words come into being along with the realities to which they relate. Eunomius' argument was that God said "let there be light" and there was light, showing that the naming of light preceded the existence of light. Gregory refutes this by asking not to take everything too literally in Scripture. When the Scripture says "Heavens declare the glory of God", it does not mean that the Heavens are using words, for the Scripture itself says "there is neither speech nor language" in the verse immediately following.[28] The real meaning of the words "And God said" cannot be understood in an anthropomorphic sense:

> "To return to the matter under discussion the word '(He) said' does not point to voice and words in God, but shows that God's will is concurrent (*sundromon*) with his power (*dunamis*), in a manner calculated to make clear to our more sense-oriented minds the true understanding of reality. Since all realities consist in God's will, and it is the custom of men to indicate the will by words and thus to harmonise work with will, the scriptural account of the origin of the cosmos is a sort of beginner's introduction to the knowledge of God, presenting the power of the divine nature through things more readily understandable (perception through the senses being a means to understanding the intelligible); for this reason, Moses, in saying that things came to be because God 'said' indicates to us the moving power of God's will, and by immediately adding: 'and it was so', shows that in God's nature there is no difference between will and energy (effecting); on the contrary teaches that in God the purposing (*noēsis*) is the principle of the effecting, not that the effecting comes later than the purposing, but the two happen together and are to be understood together—the movement of the mind (*hē tou nou kinēsis*) and the effecting power of the thing (*hē apotelestikē tou pragmatos dunamis*). There is no middle given for the understanding, between the word of choice and the thing effected; but as the kindling of the flame and the appear-

28. Ps XIX. 1-3. LXX.

> ance of night happen together, the latter co-appearing with and from the former in the same manner, the work of God's will is the hypostasis of the things coming to be, but not posterior in sequence to the will."[29]

This virtual identity, though with theoretical distinctions, between the will of God, the word of God, and the reality of all that exists, becomes the basis for Gregory's epistemology. In opposition to Eunomius' argument that (*rēmata*) are created by God along with things (*pragmata*) , Gregory argues that words are our creations to classify and organize in our knowledge the diverse elements in creation, made possible because of the order of sequence in creation:

> "Lest we understand the creation to be masterless and self-originated, he (Moses) says that it came to be from the divine nature and that it subsists in a certain order and sequence. (*taxei tini kai akolouthia sustēnai*)."[30]

What is the difference here between Stoic epistemology and Gregory's way of knowing? Our knowledge of Stoic epistemology being limited as it is, particularly in relation to thinkers of the "Middle Stoa" like Panaetius and Poseidonius, any ascription of source has to be largely conjectural.

The Stoic idea of *sumpatheia* no doubt influenced Gregory's understanding of the *taxis* and *akolouthia* in creation. But it is significant that he uses but sparingly the Stoic terminology of *sumpatheia* to denote coherence in the world. Panaetius, for example, is not interested in the relation between human reason and the logos in the world.[31] Gregory similarly does not draw the analogy between the human reason and the logos as Basil does, but he still believes that it is because there is order in the universe that knowledge is possible.

Gregory is probably more indebted to the whole Stoic tradition (which was largely appropriated for Christianity by the Alexandrian

---

29. *Contra Eunomium* Lib. II. 227. GNO vol. I, 292/2-28.
PG 45. 985 D—988 B. NPNF V, p. 272.
30. *Ibid.* 225. GNO I. 291/20-24. PG 45. 985 D. NPNF V. 272.
See also CE II. 279. GNO I. 308/24. PG 45. 1005 B. NPNF V. 278,
31. J.M. Rist, *Stoic Philosophy*; p. 189.

School), for thinking in virtue as an essential and inseparable aspect of true knowledge. Clement of Alexandria approvingly cites Poseidonius' definition of the end of life as contemplating truth and order and striving then to bring them about.[32] Panaetius saw virtue as two-fold as *theoretikē* and *praktikē*, theoretical and practical. The general close link between will and reason is rather universal in ancient Greek thought.

As for the origin of names, the controversy is very old. Proclus[33] tells us that Cratylus, Pythagoras, and Epicurus agree that names arise by nature—which is the Eunomian view also. Hermogenes, Democritus and Aristotle agree that names arise by chance, and this is generally the line taken by Gregory, though instead of chance, he finds the human *epinoia* as the creator of words.

The crucial idea in Stoic ontology was to see the universe as a strict continuum.[34] *Akolouthia, eirmos,* concatenation, or *enchainement*, is thus an integral part of the continuum theory of the universe.

3. Akolouthia and Scripture

Parallel to the *akolouthia* or the mental following up of the coherence in the order of created reality is the *akolouthia* or the following up of the Holy Scripture. In fact the scriptural *akolouthia* will manifest the cosmic *akolouthia*. In the Commentary on Ecclesiastes, Gregory says:

> "The following up of the test (*hē ephexēs akolouthia tou logou*) leads the soul to an even greater philosophy concerning created beings. For it shows that the universe is continuous (*suneches*) within itself, that there is no sort of break (*lusis*) in the harmony of created beings, but that it has co-breathing (*sumpnoia*) of all the beings with each other. The universe is never separated from this internal continuity (*tēs pros heauto sunapheias*), but all things remain in existence moved by the

32. *Stromateis* 2. 21. 129.

33. In *Cratylus*, 16. cited in H. Diels, *Die Fragmente der Vorsokratiker* (4th ed.), Berlin 1922, p. 68.

E. Tr. K. Freeman, *Ancilla to the Pre-Socratic Philosophers*, Oxford 1971, p. 97.

power of True Being (*tei tou ontōs ontos dunamei perikratoumena.*)"[35]

The sympatheia doctrine of the Stoics is more obvious in this passage, and the Editor of the Jaeger edition draws our attention to fragments of Heraclitus (ca. 500 B.C.), Hippocrates (ca. 460-380 B.C.) and Plotinus (205-270 A.D.). Heraclitus, in his aphoristic manner had said:

> "It is wise to agree that all things are one (frag. 50).
> "The hidden harmony is stronger than the visible" (frag. 54).
> "If we speak with intelligence, we must base our strength on that which is common to all, as the city (is based) on the law, and even more strongly. For all human laws are nourished by one, which is divine. For it governs as far as it wills, and is sufficient for all things, and more than enough"
> (frag. 114).

Gregory had developed these ideas in the Catechetical Oration, and ascribes it to "tradition received from the Fathers" *(para tōn paterōn diedexametha):*

> "Thus comes to be the symphony of the creation within itself (*pros heautēn sumphōnia tēs ktiseōs*), the opposition of the natures in no wise disrupting the continuity of the co-breathing (*oudamou tēs phusikēs enantiotētos ton tēs sumpnoias heirmon dialuousēs*)."[36]

But the same "hanging-together" as seen in the creation can be seen in the Scriptures also. For it is one Holy Spirit who animated the entire corpus of Old and New Testaments,[37] though this mutual

34. See Sambursky, S. *Physics of the Stoics*, New York 1959.

35. *In Ecclesiasten.* Oratio VII. GNO V, p. 406/1-7. PG 44. 724 D.

36. *Oratio.* Cat. VI. PG 45. 25 C. Srawley, p. 29. 12-30. NPNF V. 480 A. See also *De opificio Hominis* I and XIII. 7. PG 44, 129 C. PG 44, 165 A ff.

37. That the Holy Spirit is the author of the entire corpus of Scripture was a basic principle for Gregory. Besides stating so expressly on several occasions (see e.g. CE I. 225. GNO I, p. 15/15 ff. PG 45. 985) he used the expression "the God-inspired Scriptures" (*hē theopneustos graphē*) rather consistently and frequently. To take one commentary as example in the commentary on the canticles, the expression is used 17 times (GNO vol. VI. 4.16; 5.12; 2.12; 25.5; 26.17; 52.2; 76.14; 157.12; 173.5; 175.3; 263.7; 269.12; 342.9; 404.14; 436.1; 447.3; 447.14).

coherence or *akolouthia* of the Scriptures may not be at once obvious. We have to search for the coherence of the scriptural accounts behind and beyond the appearance of mutual contradiction. As he says in his epistle to his brother Peter, Bishop of Sebastia, a sort of covering letter for his Easter Gift entitled *On the Making of Man,* it is necessary to find a coherent picture of man by "fitting together" the various elements received from Scripture and from our own reasoning:

> "But even in those things regarding him (Man) that seem to be opposed to each other through the dissimilarity between what came to be in the beginning and what is now to be seen, through a kind of necessary sequence, it is necessary to weave together the explanations of Scripture and the notions discovered by our own reasonings, in order to construct one account which has consistency and order in relation to what appears to be opposed to each other, and which brings these apparent contradictions to one single end."[38]

There is, for Gregory, certain basic apparent contradictions between Scripture on the one hand and reality as we observe it on the other. Scripture says, Man is created in the Image of God, who is the perfection of all Good. But we see Man, who is more evil than good. How can man who is mortal, feeble, corporeal, finite, sinful, be the image of Him who is immortal, almighty, incorporeal, infinite and holy? Here the teaching of the Church helps us to overcome what appears like a contradiction, by drawing our attention to the whole sequence of reality in a systematic manner, i.e. what happened in the beginning, what is to take place in the end, and what is happening in between. This *akolouthia* from the beginning *(archē)* to the end *(peras)* constitutes the framework for the proper understanding of the Scriptures and of the right evaluation of secular knowledge.

But there is a certain *akolouthia* within Scripture itself, a certain order in which things are stated, which has significance for life itself. Take the five books of the Psalms, or the five divisions of the book of Psalms. The first 8 chapters of Gregory's commentary on the titles of the Psalms are devoted[39] to establishing the thesis, far-

38. *De Hominis opificio*. Preface. PG 44. 128 B. NPNF V, p. 387 B.

39. *In Inscriptiones Psalmorum*, chpt. 1-8. GNO vol. V, pp. 24-65. PG 44. 432-481.

fetched though it may seem to our modern minds, that the sequence and order of the five books correspond to five stages in the life of advancing towards the perfection of God. The 9th chapter begins thus:

> "In these five divisions of the psalms then we understand a sequence *(akolouthia)* as if a series of steps standing one above the other in a certain order which is consecutive, signifying thereby that the last sound of each division has the meaning of some sort of station or stopping place in the discourse, comprehending in itself the ends of the stages previously traversed."[40]

In other words, not only is there a consistency between Scripture and the reality of creation, but there is also a similar consistency between Scripture, cosmic order, and the life of growth into Christian perfection. The 150th psalm, for example, is the final triumph of the good when all the choirs join together in the doxology and evil has no more power.

The recital of the psalms is thus linked to the growth in perfection—an essential notion in the tradition of Christian spirituality. The first psalm refers to the beginning of the life of virtue, which means separation from the life of evil.[41] From there, by stages, the soul is led to the summit of perfection (*anabasis pros to akrotaton tou makarismou*).[42] At the end all the choirs of the creation join together in one mighty song of praise in total unison, for evil has been totally banished, and all are reconciled to God, even we who have been standing apart from the good because of our sin. It is rather easy for us to see how such a strictly allegorical interpretation of the titles of the psalms seemed natural to Gregory. The Septuagint titles clearly show that even at that early date the translators experienced difficulty in making sense of these titles. The Septuagint titles speak constantly about *tōn ālloiōthesomenōn* (the alienated or changed or estranged people) and *eis telos* (towards

40. *In Inscriptiones Psalmorum*, I. GNO V. 65/5-10. See also 67. 15 ff.
41. GNO V. 67/25-68/1.
42. *Ibid.* 68/5-6.

perfection).[43] Besides the Greek term for the titles of the Psalms is *stelographia,* which literally means inscription on a tablet of stone (*stēlē*=block of stone, monument, boundary-post), and could easily be interpreted allegorically as monuments or road-signs. In fact Gregory says explicitly that this is how he understands it; in relation to the title of Psalm 55: *Tōe David eis stelographian.* In fact the whole LXX title of Psalm 55 has considerable interest for Gregory, for it reads: "*To the goal, regarding the people who have become separated from the holy ones, for David, for Inscription (road-sign), when those belonging to other tribes overwhelmed him in Geth.*"[44] Gregory interprets these inscriptions as monuments of victory in the onward march of the soul in the conquest of evil, culminating in the final triumph-song:

> "*Eis stēlographian* should signify the word indelibly engraved on memory as if a memorial for the soul to remember the patterns of the good inscribed on a tablet. These are in a sense the victory monuments left by the captain of the forces of David (*hē tou archistratēgon tēs dunameōs tou Dabid aristeia*) according to which passion becomes double (*diploun*—divided into two) in the face of the enemies, part destroyed by fire, part obliterated through smiting."[45]

The title inscriptions of the Psalms are thus construed by Gregory, by an allegorical method that seems to us heavily strained, to be indications that the psalms denoted various stages in following the road from vice to virtue.

The main point is for us to note that there is an *arche* and a *telos,* a beginning and an end, and an ordered sequence in what is in between, at three levels—in the cosmos, in the Scriptures, and

---

43. *eis to telos* appears in the titles of the following psalms in the LXX—4, 5, 6, 8, 9, 10, 11, 12, 13, 17, 18, 19, 20, 21, 29, 30, 35, 38, 39, 40, 41, 43, 44, 45, 46, 48, 50, 51, 52, 53, 54, 55, 56, 57, 58, 59, 60, 61, 63, 64, 65, 66, 67, 68, 69, 74, 75, 76, 79, 80, 83, 84, 87, 108, 138, 139, i.e. in 56 of the 150 psalms, but it is difficult to find any sequence here. But it could still be seen as a kind of road-sign on the way to perfection. "Towards perfection". Our Hebrew MSS have (*lemenatseah*) which could mean "to the chief of the choir."

44. *Eis to telos, huper tou laou tou apo tōn hagiōn memakrummenou. Tōe David eis Stēlographian, hopote ekratesan auton hoi allophuloi en Geth.*

45. *In Inscr. Psalmorum.* II. IV. GNO V, p. 80/8-15.

in the life of perfection. In all three we have to follow the *taxis* of the *akolouthia,* the order of the sequence.

4. Akolouthia of Evil and of Good

The Christian life itself is seen by Gregory as a break with the flow of evil in which we are caught, so that we are able to leave the *akolouthia* of evil in which sin has placed us, and move along the *akolouthia* of good towards perfection. The fifth book of the Psalms speaks of the final end of human salvation.

> "Lifting up thus along with himself those who are like this, the noble prophet (David) leads us to the climbing of the fifth mountain ridge, in which is the full attainment and summing up of human salvation *(Sumplerosis te kai anakephalaiōsis tēs anthrōpinēs sōtērias)*."[46]

This is the movement from non-being to being. Gregory refers to Moses who "philosophizes about the changing and the unchanging" that "He who eternally is (God) truly is, while the one who is continually becoming (man) is not (for this is change, the transformation from being in being, to being in non-being), and showing that by the same power the nature flows down towards evil, and again through conversion (repentance) is led up towards the good since the illumination of God is capable of rekindling the life of the human being, thus revealing to us by the word the grace that has come from God."[47]

Mankind, having strayed from the city of God, which is its true dwelling, is now wandering in the desert, without God, who is true meat and drink for man. We have strayed from the way which is Christ, but in his grace, he comes to find us in the desert and puts us back into himself, for he is both way and the city of life towards which the way leads. (Ps 107: 7) But we were in the darkness, in the shadow of death, fettered by sin (Ps 107: 10), unable to move in the direction of the good, weighed down by the weight of evil (*Bareia tis esti kai katōpherēs hē kakia*).[48] It is this weight of evil that is continuously dragging us downward into the pit (*eis ton*

46. *In Inscr. Psalmorum.* I. VII. GNO V. 52/18-21. PG 44. 465-468.
47. *Ibid.*, GNO V. 52-53. PG 44. 468.

bothunon heautēs sugkathelkousa) of non-being. Referring to Psalm 107:33-34a, which says:

> "Sometimes he turned rivers into desert
> Springs of water into arid ground
> Or a fertile country into salt-flats"
>
> (Jerusalem Bible)

Gregory comments:

> "What it (Scripture) calls rivers are the flow of the passions and the springs of water (*diexodous hudatōn*) indicate the akolouthia of evil (*tas kakōn akolouthias*), Men always attaching evil to evil, extend evil into a flowing stream."[49]

Breaking the chain of evil in which we are caught, Christ places us on the road to the city of heaven, which is the *akolouthia* of the good.

The chain of evil is also spoken of in *Contra Eunomium* Book I.[50] Evil cannot be overcome with evil. Only the good can stop evil. So it would be useless for Gregory to repay Eunomius with evil, for all the evil that the latter has done to Basil.

> "Evil happenings can be counteracted only by their opposite. And evil words and actions would not have developed to such a pitch if some kind word or deed had intervened to break the continuity of the chain of evil (*to suneches tēs tōn kakōn akolouthias.*)"

Evil is a chain-reaction that leads to non-being, in which mankind is caught. Creation is a chain-reaction that should lead to being, but evil goes in the direction opposite to that of creation. The need for man is for this *akolouthia* of evil to be broken, so that he can be placed again in the *akolouthia* of the good.

Thus intellectual *akolouthia* is inseparable from moral *akolouthia*. Knowledge of the world through *akolouthia*, or knowledge of the

---

48. *Ibid.*, GNO V. 57/23. PG 44. 472 D.
49. *Ibid.*, GNO V. 61/5-8. PG 44. 477 A.
50. GNO I, p. 53/2-10. PG 45. 277 B.

Scriptures through its *akolouthia*, and the weaving together of the two into one discourse, would itself be quite useless if it does not lead to a break with this moral *akolouthia* of evil and does not become placed in the *akolouthia* of the good. This is the basic epistemological principle of Gregory's Christian philosophy.

Part Two

GOD—CREATION—MAN

CHAPTER IV

GOD AND HIS CREATION-I
Diastema - Discontinuity

Gregory was shrewd enough to see that human language was inadequate, not merely to describe God's Being, but even clearly to describe the relation between Creator and creation.

It is axiomatic for Gregory that human thought has to stop at the boundary of creation; it cannot jump the gap that separates creation and Creator. In fact the notion of the gap or *diastēma* is so central to the structure of Gregory's thought that he can apply it negatively or positively in three ways—(a) to God, negatively, (b) to the creation, positively, and (c) to the relation between God and the creation in a specially qualified sense.

1. Diastema—the Concept

The word *diastēma* normally means "standing apart", being derived from *diastēnai*, and *dihistēmi* (stand apart, set apart, or separate). The concept of *diastēma* as a musical interval between notes on a scale can be traced as far back as Archytas Tarentinus,[1] the Pythagorean Philosopher who lived in the fourth century before Christ, and to Aristoxenus Musicus,[2] his contemporary, and can be found also in Aristotle. Euclid uses the word to denote the distance from the centre to circumference (radius) of a circle. "Longinus" the literary critic uses it for distinctiveness of style in writing. In astronomy it referred to the equal distances kept between astral bodies in their movements. Temporally it could mean, in ordinary language, the distance between two events. In modern Greek it simply means "space" in such expressions as

1. Diels, *Fragmente der Vorsokratiker.* 261+262. IVth Edition Berlin 1922, p. 335, lines 5, 8 and 12.
2. See Lidell & Scott, *Greek-English Lexicon.* S.V. *diastēma.*

"space travel" or "space research". One could say that the general notion of *diastēma* refers to the distance between two points, spatial or temporal.

The Stoics were perhaps the first to use this term in a philosophical sense. Zeno and Chrysippus both used it in relation to their refutation of the cosmological theories of the atomists, but it had no central significance in their thought.[3]

In the history of Christian thought, however, *diastēma* became a key notion in the Arian controversy. The basic claim of Arianism is that "there was a then when the Son was not". If this is conceded then there is an interval between the origin of the Father and the origin of the Son, since normally the Father has to be older than the Son.

Alexander of Alexandria, the orthodox bishop in whose diocese Arius was a Presbyter, writes to Alexander of Constantinople at a very early stage in the Arian controversy, that the Father begat the only begotten Son, as eternally present to Him (i.e. to the Father), not temporally, nor out of that which is extended (*ou chronikōs oude ek diastēmatos*) nor out of that which is not.[4]

The Antiochean "Creed of the Long Lines" (*ekthesis makrostichos*, 345 A.D.) says clearly:

> "nor was a then when he was not, as some say, unscripturally and erroneously, that some sort of a time-interval (*chronikon ti diastēma*) has to be presupposed in relation to him."[5]

The document clearly says:

> "In saying that the Son is in himself, and both lives and exists like the Father, we do not on that account separate him from the Father, imagining place and interval between their union

3. See *Stoicorum Veterum Fragmenta*. Ed. H. von Arnim, Leipzig 1903. Vol. I, p. 26, Vol. II, p. 164 etc.

4. PG 18, 557 B.

5. PG 26. 729 A. Athanasius, *De Synodis*, 26; Socrates, *Hist. Eccl.* II. 19. 7-28. Eng. Tr. in J. Stevenson, ed. *Creeds, Councils and Controversies*, London, SPCK, 1972, p. 23. For a more common use of *diastēma* in St. Athanasius, see *De Incarnatione*, ch. 26, lines 15 and 29 (Oxford Text 1971). In both instances it means the time-interval between the death and resurrection of Christ.

> in the manner of bodies. For we believe that they are united with each other without mediation or interval, and that they exist inseparable."[6]

St. Basil also uses the word *diastēma* in this Trinitarian-Christological sense, and accuses the Arians "that they distinguish on the one hand the Son from the Father, and on the other the Holy Spirit from the Son, by a temporal interval."[7]

By the fifth century *adiastatos* or "unextended" had become a qualifying adjective for the Holy Trinity.[8] But Gregory finds a more comprehensive use for the term for the whole structure of the Creator-creation relationship, by positing: (a) that on the one hand there is no *diastēma* or extension or gap internally within the uncreated nature of God; time-space extension does not apply to the godhead, therefore there can be no time-space gap either *in* or *between* the three Persons of the Trinity; (b) that, on the other hand, all created existence is by nature extended in time and space, and therefore *diastēma* is the characteristic of created being; and (c) that there is an ontological and therefore epistemological gap between the Creator's being and that of the creation which the human mind or any other created mind is incapable of traversing.

2. Diastema and the Transcendence of God

It is important for us to note Gregory's originality here. He is following neither Plato nor Aristotle, neither the Stoics nor Plotinus. For the Stoics, "God is the whole world with all its parts,"[9] or sometimes "the mind of the world"[10] or "the soul of the world."[11] As Marcus Aurelius put it:

> "This thou must always bear in mind, what is the nature of the whole, and what is my nature, and how this is related to

6. *Ibid.* J. Stevenson, *op. cit.*, p. 24.

7. *"Chronikois diastēmasi tou men patros ton huion, tou de hiou to pneuma to hagion diairousi".*

De Spiritu Sancto, 59. PG 32. 117 B. See also PG 29. 596 B.

8. See e.g. Cyril of Alexandria, *Homiliae Diversae*, 4 in Aco 104.28—*tēn adiastatēn triada.*

9. Arnim, *Fragmenta* II. 258.

10. *Fragmenta* I. 157.

11. *Fragmenta* I. 532.

> that, and what kind of a part it is of what kind of a whole; and that there is no one who hinders thee from always doing and saying the things which are according to the nature of which thou art a part."[12]

The whole cosmos is God and man is both a part and a miniature version (microcosm) of the macrocosmos, for Stoic thought. As the soul is in the body, so God is in the universe. Gregory departs totally from this by positing the absolute transcendence of God as we shall see.

Neither does Gregory follow Aristotle in seeing the relation between God and the world in terms of cause and effect. Gregory does not deny the possibility of inferring God as cause from the world as effect. But this is neither sufficient for salvation, nor is it a knowledge of God's *ousia*.

> "For being by nature invisible, He (God) becomes visible only in His operations, and only when He is contemplated in the things that are external to Him. But the meaning of this Beatitude (Mathew 5:8) does not merely indicate that we can infer the nature of the cause from its operation, for in that case even the wise of this world might gain a knowledge of transcendent wisdom and power through the harmonic structure of the universe."[13]

While Gregory can accept the Aristotelian notion of God as First Cause or Uncaused Cause (which is the key to Eunomius' system), he is not satisfied with that kind of knowledge of God. Such knowledge is inferential, and is not saving knowledge. It is the kind of theology any charlatan can engage in, without any change taking place in his life. The true knowledge of God must result in and accompany a change of the very being of man, as Gregory goes

---

12. *Meditations* II. 9. Eng. Tr. by G. Long, cited in Sanbursky, *Physics of the Stoics*, p. 145.

Gregory cites the Stoic view of God in PG 46. 559 B-C. In *Contra Eunomium* III (GNO II. 106, line 3) Gregory accuses Eunomius and his followers of being Stoics and Epicureans. (NPNF V. 171 B).

13. Gregory of Nyssa, *De Beatitudinibus*, oratio vi. PG 44. 1269 A-B. Eng. Tr. in Musurillo, *From Glory to Glory*, p. 100.

on to say in this sermon on "Blessed are the pure of heart, for they shall see God."

The world is an orderly and sequential chain of causes (*akolouthia*), but God is not simply the first link of that chain. There is a *diastēma* or interval or gap between God and the causal chain of the world. Gregory can accept the conclusion of Aristotle's *Metaphysics*:

> "It is clear, then, from what I have said, that there is a substance eternal, unchangeable, and separate from sensible things. We may likewise infer that it cannot have spatial magnitude; for since nothing finite can have infinite power, it could not have that infinite power which it reveals by causing motion eternally. That rules out finite magnitude. And it cannot have infinite magnitude, because there is no such thing. It must also be free from change of quality, for all the other sorts of change presuppose locomotion."[14]

But both Aristotle and the Neoplatonists assume some sort of continuity between the world and God. Of course it is not spatial continuity, since even for Aristotle, the supreme Intellect or Prime Mover has no magnitude and therefore cannot be spatial.

Thus the idea of a *diastēma* between God and His creation, totally impossible for the Stoics, is practically inconceivable for Aristotle or Plotinus. For Aristotle, the Prime Mover or First Cause, is also the Final Cause, towards which everything moves.

> "The final cause, then, moves by being loved, while all other things that move do so by being moved. Now that which is moved is capable of being otherwise than it is, so that if its actual mode of existence is the primary kind of local movement, then, in so far as it is subject to change, it is capable of being otherwise in respect of place even if not in respect of substance. The unmoved mover, on the other hand, has no contingency: it is not subject even to the minimal change (spatial motion in a circle), since this is what it originates. It exists, therefore, of necessity; its being is therefore good, and

14. Aristotle, *Metaphysics*, Book Lambda: VII. Eng. Tr. John Warrington, Everyman's Library ed., 1970, p. 347.

> it is in this way that it is a principle of motion. On such a principle, then, the whole physical universe depends."[15]

So far Gregory would agree with Aristotle. But Aristotle does not go far enough either in the direction of total transcendence, or even of temporal transcendence, which may be implied in what he says, but is not definitely worked out.

For Aristotle, wisdom is the knowledge of first causes, and that is the *skopos* or intention of his metaphysics. The more universal one's knowledge, the wiser he is.[16]

> "What are most knowable are first principles and causes, for it is through and from these that other things are known, and not they through the particulars falling under them. The most authoritative science, reigning supreme over the subsidiary, is that which knows for what purposes every act takes place, i.e. the final cause, the good in each particular instance, and in general, the *summum bonum* in nature as a whole."[17]

For Aristotle thus the First and Final Cause, the Supreme Good, is well within the range of human knowledge, and there is no *diastēma* or gap between the knowing mind and the Supreme Good, either epistemologically or ontologically. For Aristotle "Philosophy", the knowledge of "being qua being,"[18] is primarily an intellectual quest, for the object of the particular scientific disciplinec alled philosophy or metaphysics. The quest for this knowledge of being qua being proceeds, for Aristotle, through the principles or the Laws of Contradiction and Excluded Middle, and not, as for Gregory, through the life of righteousness or virtue.

Plotinus also assumes basic continuity between God and His creation. Plato had left the problem unresolved explicitly, but God for Plato belongs integrally to the true world of ideas, which can be known through mathematical and logical knowledge as distinct from logical knowledge.[19] The intelligible world for Plato is a

---

15. *Metaphysics*: VII. Eng. Tr. Warrington, *op. cit.*, pp. 345-346.
16. *Metaphysics* A: II. Eng. Tr., p. 54.
17. *Ibid.*, pp. 54-55.
18. *Metaphysics* Bk. K. III. Warrington, p. 119.
19. See *Timaeus*. 28, 29.

living being, perfect, eternal, and the only real universe, which has true being. The *kosmos noētos* or the intelligible universe thus becomes a Second God,[20] and constitutes also the soul of the body which is the sensible world, forming its principle of movement as well as of intelligibility. The intelligible world and the sensible world together constitute one universe and not two[21]—one universe comprised of "on the one hand an intelligible and unchanging model, and on the other a visible and changing copy of it."[22] Plato in fact adds a third element to this universe namely "the receptacle"[23] for the world of becoming, called space.

But it is not clear whether this "second God" who is the universe is continuous with the first God or not. This problem of the relation between the One and the Nous was resolved only in Alexandria, by Philo and by Plotinus in two different ways. Philo resolves the question by saying that the Alone was Alone before the creation of the World of Ideas. The First Cause is beyond all virtue and knowledge, and even beyond the good and the beautiful.[24] Here clearly we are in the realm of the Hebrew Scriptures, and Philo is here interpreting not Plato, but Moses.

Plotinus rejects the Philonian solution and advances his own. In the fifth Ennead, he puts it clearly[25] :

> "It is necessary then, that if the One existing unmoved, has also a second with it, the former cannot subsist inclined to-

20. *Timaeus*. 34b.
21. *Ibid*., 31b.
22. *Ibid*., 48e, 49a.
23. *Ibid*., 49, 50, 51.
24. Philo, *De opificio Mundi*. 8. 23.
25. *Enneads* V. 1. 6. My translation is different from that of Stephen Mackenna. See *Plotinus, The Enneads*, Faber, London 1969, p. 374. I give below this translation, which seems to me to take some liberties with the text :

> "Given this immobility in the supreme, it can neither have yielded assent nor uttered decree nor stirred in any way towards the existence of a secondary. What happened then? What are we do to conceive as rising in the neighbourhood of that immobility? It must be a circumradiation—produced from the supreme but from the supreme unaltering—and may be compared to the brilliant light encircling the sun and ceaselessly generated from that unchanging substance. All existences, as long as they retain their character, produce—about themselves, from their essence, in virtue of

> wards the latter, nor desiring it nor moved towards it. How then ? How are we to conceive the latter remaining around the former ? As an emanation (effulgence, radiance, *perilampsis ex autou*) from it, but remaining dependent on it, as the Sun holds around it the light radiating from it, which is eternally generated by it. And, all things that have existence, so long as they exist, by necessity, give out, around them, from their essence towards their outside, from the power present in them, an outward-moving hypostasis, being something of an image of the archetype from which it flows out, fire, for example, emitting heat around it. And snow does not keep its coolness to itself. The best examples are fragrant objects, for as long as they exist, something flows out from them and is perceived wherever they are present."

Plotinus' solution to the vexing problem between the One and the Many is a relatively simple one—emanation, circumradiation, effulgence, a property not peculiar to God, but equally shared by all existing beings. For Plotinus God and the cosmos are totally continuous, like the Sun and its light, like fire and its warmth, like snow and cold. All being, God and the universe, is in one continuous chain. The "Second Existence" of Plotinus, the *nous*, is the same as the "intelligible world" of Plato.

We see clearly a difference of approach here between the Jewish Philo and the Indo-Hellenic Plotinus.

At this point, Gregory of Nyssa, faithful to the Judaeo-Christian tradition, rejects the Indo-Hellenic theory of continuity between the One and the Many and adopts the Jewish solution of discontinuity. Among the various hellenic solutions—of Plato, Aristotle, and the Stoics, Gregory can choose none. It is Moses that is here authority for Gregory as for Philo.

One cannot agree here with Father Leys[26] who states that Gregory follows Plato rather than Aristotle in his conception of the

---

the power which must be in them—some necessary, outward-facing hypothesis continuously attached to them and representing in image the engendering archetypes; thus fire gives out its heat; snow is cold not merely to itself; fragrant substances are a notable instance; for as long as they last, something is diffused from them and preceived wherever they are present."

God-world relationship. The concept of *diastēma* between God and His creation which is the basis of the transcendence of God, as a specific Judaeo-Christian concept, is the controlling category for Gregory—not Plato or Aristotle, certainly not the Stoics or the Neo-platonists.

3. Diastema as Extension

We have no English word by which to translate *diastēma.* To translate it as gap or interval could be to miss out its meaning of extendedness. So we will use this Greek word in English, rather than use distance, distension or interval, and explain every time that it has another meaning as well.

The use of the concept has been studied in some detail by Paul Dandelot,[27] who leaves the word untranslated, and says "one is tempted to translate as space, time and measure, as the context seems to indicate. What it means in fact is all the characteristics that the intelligence apprehends in created existence, but which are recognized to be absent in God."[28] Balthasar and Danielou translate it into French as *espacement,* but this is also a limited sense.

Gregory's concept of *diastēma* seems so daringly original, that we should be careful not to assimilate it into the categories of other thinkers.

For Gregory, *diastēma* is intimately connected with movement and change. A translation like "standing apart" gives the impression of being static, while for Gregory it is impossible for the creation to be static except in death and non-being. All is in movement, all is changing—that is the very nature of creation.

The creation is an orderly process in space and time—a *taxis kai akolouthia, diastēmatikōs ek tinos eis ti tēi zoēi diodeuousa* (an order and sequence, distensionally journeying through life from something to something).[29] It is a movement from origin to perfection, from *archē* to *telos,* and an *hodos* to traverse from beginning to end.

It is thus extension not only in space, but also in time. For what is time after all? It is the interval between the beginning and

---

26. R. Leys, *L'Image de Dieu*, p. 39.

27. P. Dandelot, *La Doctrine du Diastéma chez Saint Grégoire de Nysse*, Dissert. Louvain 1960. Typescript, 240 pp.

28. *op. cit.*, p. 83.

29. *Contra Eunomium* II. GNO I. 246/27. PG 45. 933 B.

the end or between inception and perfection. This latter is a particular characteristic of creation, totally absent in God. Gregory never tires of making this distinction in the dispute against Eunomius, for the heart of Eunomius' contention is that the Son has a beginning and therefore that he cannot be God. From the beginning Gregory has to attack this contention that the Second Person of the Trinity *has* a beginning, and to state the teaching of the Church that He *is* the beginning of all things. It is with this intention that he makes the clear line of demarcation between created nature and uncreated nature.

The argument goes somewhat like this. If the Son has a beginning in time, and if there is an interval of time between the Father and the Son, then it is logically inevitable that the Father also had a beginning in time. The very sophistic argument of Gregory can be analysed thus: what are the steps of refuting Eunomius' argument?

Eunomius' *Premiss A* : The Son has a beginning in time.

Eunomius' *Premiss B* : There is a *diastēma* between the Father and the Son.

Argument : Now, *diastēma* is always a finite distance from one point of space or time to another.

So, logical *conclusion*: The beginning of the Son, plus the time of the interval, gives the time of the beginning of the Father. Therefore, the Father also has a beginning. Therefore, if the Son, who because he has a beginning, is not God, then the Father who also has a beginning is not God, which is absurd.

Final conclusion : Therefore the Son has no beginning and there is no *diastēma* between the Father and the Son.[30]

The argument, as we have said, is clever, typical of the Second Sophistique, but Gregory is shrewd enough to anticipate a counter-argument,[31] showing that Gregory's position is also absurd, by the same token.

Gregory's Premiss A : The creation has a beginning and is not eternal.

Gregory's Premiss B : There is a *diastēma* between the Creator and the creation.

30. *Contra Eunomium* I. 344-358. GNO I. 128/30-133/7. PG 45. 357 D-364 A. NPNF V. 67 B-68 B.

31. *Ibid.*, I. 359. GNO I. 133/8ff.

Now, given *diastēma* is finite, as in the argument against Eunomius, the beginning of the creation plus the *diastēma* gives the beginning of the Creator. So the Creator has a beginning and is not God, which is absurd. So Premisses A and B cannot stand.

4. **Diastema as Gap**

In order then at the same time to deny *diastēma* between the Three Persons of the Trinity, and to affirm *diastēma* between the Creator and the creation, Gregory has to lay down an axiom—the total incommensurability of Creator and creation. In the process he also destroys all possibility of *analogia entis*, i.e. he denies that the Being of God can be known by magnifying infinitely what we experience of God in the creation. The distinction between the Creator's nature and that of the creation is clearly laid out at several points. We cite here first one of the most extended of these many passages:

> "Since it is confessed in all the expressions of opinion of those who rightly worship, that of all beings, some are through creation while the other is before creation, and that while the divine nature is believed to be uncreated, (in which, however, there is distinction between the cause and that which undiastatically (*adiastatōs*) has its hypostasis from the Cause as the true faith teaches) the created nature is understood as dimensionally extended (*en parastasei tini diastēmatikēi*); all chronological order and sequence of events can be apprehended only in relation to the ages; the nature that is 'before the ages' transcends all distinctions of elder and younger, since reason cannot understand the divine and blessed Life (of God) in the same terms as it specially observes in created existence. The whole creation, as we have said, having come into existence according to a particular sequence of order (*kata tina taxeōs akolouthian gegenēmenē*) can be measured in terms of the dimensions of the ages (*tōe tōn aiōnōn diastēmati parametrcitai.*) So much so that if one ascends logically following the sequence of created beings to the beginning of all events, the search shall end at the foundation of the ages. The Being (*ousia*) which is beyond the creation is far removed from all dimensional conception (*pantos diastēmatikou noēmatos kechōrismenē* and eludes all temporal sequence; it escapes and stops

all attempt to find the *arché* from which it takes its origin or the *peras* (end) towards which it advances by any sort of change or ordered sequence. To our reasonings, which traverse the ages and all that have taken place in them as if an immense ocean, the understanding of divine nature gives no sign of its own knowable beginning, yields no 'direct apprehension' (*kataleptikē phantasia* a Stoic term meaning clear immediate vision) to the mind that stretches itself to apprehend. So that anyone who is inquisitive about the 'seniority of the ages' and wants to ascend to the source of all being in his mind, can get no foothold for his thoughts, the object of search will be infinitely advancing and will offer him no stopping place for his inquisitiveness.

"The matter should thus be clear to anyone with even a moderate insight into the nature of beings, that there is no measure for the divine and blessed Life; for it is not in time, but time is from it; the creation, on the other hand, is carried forward, from its acknowledged beginning towards its own goal, journeying through temporal dimensions; so that it is possible as Solomon says somewhere (Wisdom 7:18) to discern a beginning, an end, and a middle for it (*archēn kai telos kai mesotēta*), and to signify the sequence of temporal events through divisions of time. But the Transcendent and Blessed Life has no measure or space, no dimensional traversing in creation. All things that have come to be, confined within their own measures assigned in accordance with the wisdom and good pleasure of their Creator, are measurable because bounded, in accordance with the needs of mutual harmony. So even though the weakness of the human mind cannot go all the way in the understanding of things in creation, it does not doubt that all things are bounded by the power of their maker, and are within the bounds of created existence. But this creative power which sets the limits for all created beings, has itself no circumscribing limit or boundary, it shuts out all conception striving to ascend to the origin of the Divine Life; it eludes and goes beyond all futile investigation and inquisitive and ambitious striving to get to the boundary of the unbounded. All effort of discursive reasoning to go beyond the ages and to ascend above the *diastēma* can only climb to the point where it sees that the gap cannot be traversed; the measure and limit, so to speak, of the movement of human reasonings and operations seems to be time (*ho aiōn*) and all that is contained in it. That which transcends these remains unattainable, and inaccessible, unsullied by any object capable of human apprehension. In it is there neither form (*eidos*) nor place (*topos*), neither magnitude (*megethos*) nor the measure of time (*to ek tou chronou metron*), nor anything else that can be apprehended. So, by necessity, the comprehending power of the human mind, seeking a foothold for thought in some object in the ages and in creation, must fall back from every side of this incomprehensible nature, and stay within its own kindred and congenial realm.

"As I said, it should be understandable to all with however moderate an insight into the nature of beings, that the Creator of the whole laid the ages and the space in them as a sort of receptacle capable of receiving all that came to be through creation, and in them (i.e. in space and time) created the universe. For it is not possible for anything that came to be or comes to be to have its existence except in time and space. But the nature that is self-sufficient, eternal, and all-comprehending is neither in space nor in time; but is before and beyond these, in an ineffable way, itself from itself, understood only by faith, incommensurate with the ages, unmeasurable in terms of time, subsisting from itself, at rest in itself, not divisible into past and future; for there is nothing beside itself or outside itself, by the passage of which something becomes past and something future for it. These (temporal) experiences are peculiar to those things in creation, which divide their life, by the divisions of time, into hope and memory. In that exalted and blessed power to which all things are simultaneously and instantaneously present, the past and the future come within instant all-embracing view of the Power that empowers all existent beings.

"This then is the Being, in whom are all things, as the Apostle says, constituted, and by participating in whom each of us existent beings live and move and have our being; the Being which is above all beginning, presents no signs of its own nature, known only in its impossibility to be grasped. That is

> its peculiarly knowable characteristic, that its nature is above all conceivable characteristic."[32].

Obviously it is in the Eunomian Controversy that it became clear to Gregory that because of a fundamental ontological difference between the Creator and the creation, the logic that applies to the creation cannot apply to the Creator. In the creation, father has to be senior to the son in age. This was Eunomius' key argument. Gregory could refute it only by making clear the distinction between the true, non-temporal-spatial, unextended, self-generated life and Being of the Creator on the one hand, and the contingent, temporal spatial, extended, other-generated life and being of creatures. That distinction is central to Gregory's thought, and is repeated several times in his writings. We shall cite another significant and clear passage, from the same corpus of writings against Eunomius:

> "Huge and untraversible is the gap which separates the created nature from the Being of the Creator. The former is bounded, the latter has no limits. The former can be comprehended, in its own measures which accord with the good pleasure of their maker, for the latter, the measure is infinity. The former spreads itself extended in certain dimensions, and can be investigated in terms of space and time, the latter exceeds all dimensional or extensional understanding. It is like someone trying to comprehend the (human) mind which eludes the grasp of the inquisitive mind. In this life, the beginning of beings and their end can be conceived; but the Blessedness which is beyond the creation admits no conception of either beginning or end, but has eternally subsisted as he is, beyond the ideas of beginning or end, dependent only on himself, not journeying through life from somewhere to somewhere else in dimensional existence (*ou diastematikōs ek tinos eis ti tēi zoēi diodeuousa*), for it is not by participation in some other life that it comes into life, which would be logically necessary if we are to conceive a limit and a beginning to it; but it is what it is—life energising and energised from within itself, not increasing or decreasing by addition or diminution. For increase towards that which is greater has no place in that which has no

32. *Contra Eunomium* I. 36-373. GNO I, p. 133/27 to 137/8. PG 45. 364 C-368 B. NPNF V. 69 A-70 A.

> limits, and as for decrease, this cannot be conceived in the nature which cannot suffer and is incapable of diminution."[33]

The context here is especially worth noting. It is not because Gregory has a lower estimate of man that he thinks man cannot understand the *ousia* of God. No, he has so high a conception of man, that it is hardly possible to go beyond the standard he has set. But for Gregory, even the angels are in no better situation, despite their being closer to God. In a phrase hardly possible for medieval western thought which put the angels as some sort of intermediary beings between God and Man, Gregory says :

> "The power of the angels appears in every way superior compared to ours, for they can pursue the knowledge of the higher world with naked and unveiled knowing power, undisturbed by attraction to the sensible world. But, when compared with the greatness of Him who really is, perhaps if one dares to say that even their power of apprehension is not so very superior to our own feebleness of understanding, he is not being over-presumptuous."[34]

So it is not a question of the frailty of the human mind that constitutes the gap. The gap applies even to the Cherubim and the Seraphim. This fundamental division between the Creator and the creation is no Indo-Hellenic notion. It comes beyond a shadow of doubt, from the Judaeo-Christian tradition. But even Philo does not use the notion of *diastéma* in this way as a basis for the absolute transcendence of God.

This world-view of Gregory's is so central to his thought, that it occurs in practically all his philosophical writings.[35] One finds it wherever Gregory seeks to lift up the mind of man to God—whether in the commentary *On the Lord's Prayer*[36] or in the sermon *On Infants who die prematurely*,[37] or in the *Great Catechism*.[38]

---

33. *Contra Eunomium* II. 79. GNO I, p. 246/14-247/4. PG 45. 933 A-B. NPNF V. 257 A.

34. See just before the passage cited in note 33.

35. In the *Contra Eunomium* itself see GNO I. 105/19ff. [PG 45. 333 B ff) I. 113/20ff. (PG 45. 341 C ff)].

36. *In Oratio Dominica*. PG 44. 1165 B.

37. *De Infantibus*. PG 46. 173 A-B.

38. *Oratio Catech*. 6. PG 45. 25 B ff.

But in several of these passages there appear also another division of the cosmos into visible and invisible or sensible and intelligible. But in this latter division between—let us call them—the human world and the angelic world, there is no *diastēma*, though they substantially differ from each other.[39] Heaven and earth, if we may use another terminology for the two divisions of the cosmos, interpenetrate. There is no gap or abyss between them that one cannot pass from the one to the other. On the contrary human beings participate simultaneously in both worlds. We shall cite references and make clear the difference between the two parts of the universe later on.

Here the point is to note the unique nature of *diastēma* in its three aspects—(a) the absence of *diastēma* in the ousia of the Creator, (b) *diastēma* as the invarying character of created existence, and (c) *diastēma* as unbridgeable gap between the Creator and His creation.

5. The Three Aspects of Diastema

(a) *Absence of diastēma in the Creator* is the primary ground of his Incomprehensibility, and even of His transcendence. Gregory has a very modern epistemology, namely that all human knowledge has to come through the senses. Plato too concedes sense-knowledge, but he basically distrusts it. For Gregory, on the other hand, it is sense-knowledge upon which all other knowledge it built.[40] Here he has certainly been influenced by both Aristotle and the Stoics in his revision of Plato's view. Perhaps also by Strato of Lampsacus, head of the Peripatetic school from 287 to 269 B.C., who denied all distinction between reason and sense.[41] Gregory has a high, rather than low, evaluation of sense-perception as a precious gift of God.[42] But neither sense-knowledge nor the *epinoia* or

39. It is unfortunate that the English translation oı the Great Catechism (NPNF V. 480 A) gives the false impression that there is such an "interval". Gregory does not use *diastēma* to denote the difference between the sensible and intelligible world, but the word *diairesis* which means *distinction*, not *interval.* See PG 45. 25 B.

40. See E. Von Ivanka, *Plato Christianus.*

Contra Eunomium. GNO I. 107ff. PG 45. 341 B-C.

41. See Weiswurm, *The Nature of Human Knowledge according to St. Gregory of Nyssa.* Washington 1952, pp. 83ff.

42. See *Contra Eunomium.* GNO II. 328. PG 45. 1045 A.

conceptual discoveries of truth based on sense-knowledge, can lead to the knowledge of God's *ousia* for God is not an object that makes an impression on the senses. The non-extended nature of God therefore eludes all sense-perception and all conceptual apprehension. God is not an "object" in the external, extended universe, open to our senses.[41] The *theōria katalēptikē* or comprehending vision, cannot grasp the *ousia* of God, because only objects in the diastematic universe of space and time can yield the *phantasia kataleptikē*—the direct appearance to the mind—which is the basis or "category" for the *theoria katalēptikē*. The creation is *diastéma*, and the diastematic creation cannot go outside the bounds of itself by the operations of its mind.

The absence of *diastēma* in God has also much to do with Gregory's Triadology and Christology. But we cannot discuss these here.

(b) *Creation as diastēma*. On the other hand, as we have said many times, the whole creation is *diastēma* or extendedness in time and space. This creation is open to the knowledge of man. But everything in it is in movement and change; nothing in it is abiding, permanent, or unchanging. It is a constant flow. There are two kinds of movements—first, from place to place, and second, internal to each thing. Both these movements take place in the flow of time. Gregory cites the planets as an illustration of the first kind of change, and the human body as example of internal change.

The elements of movement (*kinēsis*) and change (*tropē*) thus constitute major areas of difference between created and uncreated nature. In the created nature everything is movement, change, alteration, in that of the Creator there is rest, changelessness, identity. Gregory says in *On the Making of Man*:

> "Movement (*kinēsis*) is not conceived only as change of place (*topikē metastasis*) but also to be understood as change (*tropē*) and alteration (*alloiōsis*), the immovable nature, on the other hand, does not admit any movement of alteration...that which is inclined to move or change, would no longer be divine nature...Instability and change cannot be ascribed to the Divine nature."[44]

---

43. See J. Bayer, *Gregors von Nyssa Gottesbegriff.* Giessen 1935, pp. 38ff.

44. *De opificio hominis*. 1. PG 44. 129 C to 132 A. NPNF V. 389 2.

The Stoic roots of Gregory's notion of change is discussed by Danielou, *L'être et le temps*, pp. 97ff.

The cyclic motion of the four elements that constitute the universe one into the other is a specifically Stoic notion—one could say a "scientific notion" in Gregory's time. Even the idea that all the material creation is involved in continuous change can be traced back to Chrysippus or Panaetius. Had not Marcus Aurelius said: "Everything changes" (*panta tropai*)?[45]

But Gregory carries the notion much further in linking created existence and change inseparably. For the Stoics changeability is a characteristic only of the sensible world. For Gregory, change applies even to the intelligible world. To be created is to come into being by a process of change—from "non-being" into "becoming", and that which had its origin in change continues its existence only by change.

Thus *diastēma* and *tropē* are inseparable as characteristics of created existence. From non-being all existents in creation have come and towards non-being they constantly move: "All beings which have a beginning and an end for their existence begin from non-being and terminate in non-being."[46] Diastematic existence is thus never static, but changing, change and impermanence are thus symptoms of createdness. Being dies every moment. No created being has being in itself.

> "But the things which come under our comprehension (*katalēpsis*) are such as can be understood only as extended in some diastematic manner (*en diastēmatikēi tini paratasei theōreisthai*) or conceived as locally spaced, or appear to our perspective as enclosed within a beginning and an end, since the existence of each being is bounded on both sides by non-being."[47]

In fact death occurs every moment. There is no present, only that fleeting moment which is the point of intersection between a future that is not yet and a past that is no more, that which is beyond moving into that which is behind.

---

45. *Meditations*. VII. 2. cited by Danielou, *op. cit.*, p. 98.

46. *Pan gar to archēn te kai teleuten echon tou einai ek tou mēontos archetai kai eis to mē on katalēgei.*

Contra Eunomium. II. 578. GNO I, p. 395/9-11. PG 45. 1104 D.

47. *Contra Eunomium* II. 578. GNO I. 395/3-9. PG 45. 1104 D.

"Once death was mingled with his (man's) nature, mortality was passed on to all the generations of his children. Hence we are born into a life of death, for in a certain sense, our very life has died (*nekros hēmas diedeyato bios autēs tropon tina tēs zoēs hēmōn apothanousēs*)."[48]

Even more eloquent is the description of this dead life in the dialogue *On the Soul and the Resurrection*:

"Who does not know that human nature resembles a stream, from birth to death ever advancing as by an irresistible movement, and that when that movement ceases, then comes also the end of existence. This movement (*kinēsis*), however, is no mere displacement from one locality to another (for how can the nature go out of itself?), it is, rather, an advance by alteration. And alteration, so long as it is that, can never remain at the same stage (for how can the altered remain the same?). But like the flame on the wick, which appears to remain always the same (for its unbroken continuity gives the impression of being a self-identical unit), but in fact is always wholly passing away and never remains the same (for the moisture which is drawn up by the heat is constantly burned up and transformed into smoke and bursts into flame, effecting the movement of the flame by this altering power, the substratum or fuel transforming itself into smoke-flame), so much so that anyone touching the flame twice does not both times touch the same flame (so quick is the alteration that it does not remain till the second touch); the flame is ever new and being renewed, it passes away every moment without remaining the same and is generated anew every moment. So is also the situation with the nature of our bodies. The influx of our nature and its efflux through the process of alteration of movement goes on permanently, and when the movement stops, life also ceases. So long as it is alive, there is no stopping. On the one hand, it is being filled, on the other it is breathing out, the combination of the two processes keeps it constantly moving."[49]

48. *In Canticum Canticorum*. XII. GNO VI, p. 351/2-5. PG 44. 1021 D. Eng. Tr. Musurillo, *From Glory to Glory*, p. 259.

49. *De anima et Resurrectione*. PG 46. 141 A-C. NPNF V. 462 B to 463 A.

Diastematic existence, or created existence, is constant change—a sequence of unbroken change. Change is the ontological basis of the creation's existence, for it has come to be, by change from that which is not. This applies to things as much as to men, nay even the angels are not free from the necessity to change constantly, in so far as they are created, and have come to be out of nothing. The principle applies to all creation and is expressed clearly in the catechetical oration, and affirmed, not as a personal philosophical speculation, but as a teaching handed down by tradition from the Fathers:

> "The teaching we have received by tradition from the Fathers (*para tōn Paterōn diedexametha*) goes like this: and the teaching is not some kind of a mythical narrative, but one that by our very nature we are led to believe. The apprehension (*katanoēsis*) of all existing things is a two-fold one, understanding (*theōria*) being divided into those of intelligibles and of sensibles. There is nothing among the natures of existents which falls outside this two-fold distinction. But these two have a great difference between them, so that the sensible cannot be included in the intelligible, nor vice versa, but each have mutually opposed characteristics. The intelligible nature is something incorporeal, impalpable, formless. The sensible nature, on the other hand, as its very name indicates, is bounded by the perception that comes through sense organs. But in this very sensible universe, despite the seemingly considerable mutual opposition between the various elements which constitute it, we can also discern above this conflict a harmony of opposites devised by the All-constituting Wisdom; thus there comes to be a mutual symphony (*sumphonia*) of the creation within itself, the mutual opposition of the natures never managing to disrupt the chain of conspiration (*tōn tēs sumpnoias heirmon*). In the same way, between the sensible and the intelligible also there takes place a sort of commixture and blending together by the divine wisdom, so that all may share in the good in an equitable way, and that none may be deprived of a share in the superior nature. Thus while on the one hand the appropriate "place" for the intelligible nature is the subtle (*leptē* and easy-moving (*eukinētos*) essence, the supra-mundane section (of the creation) having many characteristics with close

> affinity to the nature of the intelligible, there occurs, on the other hand, by the foresight of the Supreme One, some sort of a commixture (*sunanakrasis*) of the intelligible with the sensible creation, so that nothing in creation may be set aside (as valueless), as the Apostle says, deprived of participation in the Divine. For this reason, in human nature, composed of the intelligible and the sensible, there is a mixture (*migma*) with the divine nature, as the word about the creation of the cosmos instructs us. For, God, it says, taking the dust of the earth, fashioned man, and implanted life in it by inbreathing into that which was fashioned, so that the earthly may be elevated together to the divine (*sunepartheiē toē theioē to gēinon*) and that the divine grace may pervade the whole universe as one unit of equal worth, the lower nature being commingled with the supra-mundane."[50]

Here it is presented as official teaching that the world open to the senses and the intelligible world are not two disparate realities, but one single unit; but the whole of this single unit is separated from the Creator's nature by a *diastema* which cannot be traversed from the creation's side. And the whole of creation is subject to change, including the angelic beings, as Gregory goes on to say in the same chapter 6 of the Catechetical Oration:

> "......the Uncreated Nature is incapable of admitting the movement of change and transformation and alteration, while all that has come into existence by creation has affinity with alteration, since the very existence of the creation began from alteration, that which was not having been transformed by the power of God into that which exists."[51]

So, ontologically speaking the fundamental difference between the Uncreated Nature and the Created Nature may be summarized thus:

(1) The Creator's being is self-derived, self-generating, self-subsistent, not dependent on aught else. The creation has

50. *Oratio Catechetica*. 6. Srawley, p. 28/15 to p. 31/11. PG 45. 25 B—28 A. NPNF V. 480.

51. *Ibid*. Srawley, p. 33/15 to 34/6. PG 45. 28 C-D. NPNF V. 481 A.

no being in itself. Its being is derived from God, subsists only by the will of God, and cannot exist in itself, but is constantly and every moment recreated by God's will, on which it is dependent.

(2) God's being is perfectly good, and the perfection of all good. It is good not by participation in other good, but is good in itself. There is nothing better into which God can grow, nor does he ever become less than the perfection of all good. He is thus never in need of change. Whereas the creation is good only because the will that created it is good. It remains good only by participation in that will. It is created diastematically, i.e. it has to traverse a course from inception (*archē*) to perfection (*telos*), a process which is unceasing and infinite, since it will always be moving towards the infinite perfection of God by participation in His energies, but infinity can never be achieved by adding to finitude. So change is the eternal characteristic of creation—eternal change either towards good and the perfection of all good, which is true life, or change towards the opposite of the good, or towards non-being or death.

(c) *Diastéma as basis of incomprehensibility.* Gregory never tires of saying that there is no faculty in human nature adequate for the full comprehension of the divine essence. He does not say, as some scholars seem to suggest, that while God is intellectually not to be comprehended, man has some kind of a "mystical" faculty by which he can apprehend God—some kind of existential encounter with the ground of our existence. Practically all writers on Gregory,[52] with varying degrees of competence in the history of Christian thought, experience and spirituality, give credit to him as the first systematic expounder of Christian mysticism. Johannes Bayer tries to correct this one-sidedness by calling it a more existential mysticism—as a search of the whole man for God, in a sort of existential ecstasy:

"If the pure intellectual apprehension of God is closed for

52. Just to cite a few names: Danielou, *Platonisme*; Volker, *Gregor von Nyssa als Mystiker*; J. Maréchal, *Etudes sur la psychologie des mystiques*, Paris; Desclée de Brouwer, 1937; Diekamp, *Gotteslehre*; Bayer, *Gottesbegriff*; H. Koch, *Das mystische Schauen beim hl. Gregor von Nyssa* (Theol. Quart. LXXX (1898), pp. 397-420.

man there remains to him the prospect of a far greater union and oneness with God which involves his whole existence."[53]

The attempt to isolate this so-called "mystical-intuitive" knowledge of God from the total Christian life of worship and knowledge and faith and practice and sacramental initiation, and to make it an alternative to the intellectual knowledge of God seems to misrepresent Gregory's thought, by easy assimilation to Plotinian categories. Nor would it do to understand Gregory's theory of knowledge as a transition from Platonic to Aristotelian categories, as a neo-scholastic like Weiswurm is tempted to do, when he concludes his work on *The Nature of Human Knowledge According to Gregory of Nyssa* with the astounding comment:

> "Although the Saint's (i.e. Gregory's) theory of knowledge is not completely free from Platonic elements, it, nevertheless, presents, as a whole, a decisive step in the direction of the Aristotelian conception of knowledge. However, the Aristotelian characteristics are neither numerous enough, nor so exclusively Aristotelian, as to warrant the classification of St. Gregory's theory of knowledge as outright Aristotelian. It remained rudimentary in many respects, vague and confusing in others. But it must be regarded as an early though faltering step towards that great synthesis of Platonic and Aristotelian elements which reached its completion at the hands of St. Thomas Aquinas."[54]

Gregory of Nyssa has to be understood in his own terms, and in terms of the intellectual and spiritual context in which he lived, of which Aristotelian and Platonic categories are only as important as the Stoic and the Plotinian, the Philonic and the Pythagorean. His insistence on the incomprehensibility of God has some Platonic elements in it, but it is more of a deep conviction born out of the struggle to know everything. There is an element of personal reminiscence in that eloquent passage of Gregory's in the Seventh Homily on Ecclesiastes:

> "He who is believed to be above all that exists, is also

53. *Gottesbegriff*, p. 55.
54. *op. cit.*, p. 225.

assuredly above words. To the one who reaches out to grasp the unbounded by word or concept, he who is above all being yields no ground, and casts out the very concept which tries to say thus and so, this much and in this way. He does not know that only in being believed to be beyond all knowledge, the true conception about Him who truly is *(peri tou ontōs ontos)* is conserved. Why? For all that is in creation looks by nature to that which belongs to the same kind, and nothing which has existence can remain in existence if it steps out of itself, not fire in water, nor water in fire, not the dry in the abyss, not the wet in the dry, not in the ether the earthly (heavy) and not the ethereal in the heavy, but each element by remaining within its own natural bounds remains what it is only so long as it remains within the bounds of its own nature. If it steps out of itself, it steps out of existence itself. And the power of any sense organ is unable, while remaining within its own nature, to be transformed into that of its neighbour, the eye cannot do the work of hearing, the sense of touch cannot speak, the sense of hearing cannot taste, the tongue cannot perform the function of seeing or hearing, each has a boundary for its power to function according to its own nature. So also the whole creation is not able to go out of itself through its comprehending vision (*dia tēs katalēptikēs theōrias*), but ever remains enclosed within itself, and whatever it sees, it sees only itself. And even when it thinks it sees something beyond itself, it does not have the natural capacity to see something outside of itself. It may strive to go beyond the diastematic conception of the understanding of existents, but it never achieves it. In all discovered conceptions, it also co-apprehends always the diastēma which is inherent in the hypostasis of the conceived (object). This diastēma is nothing but the creation itself (*to de diastēma ouden allo ē ktisis estin*). For this the good, which we have learned to seek and to cherish, and with which we wish to achieve contact and to remain attached, being above creation, is also above comprehension. For how can our understanding, traversing through diastēmatic extension, comprehend the unextended nature? The enquiry, proceeding through temporal sequence by analysis, goes on to the antecedents of that which has been discovered. Even if diligent search traverses through all that is known, it discovers no mechanism by which to traverse the very conception of time (*aiōn*) itself, being unable to stand outside of itself and to surpass time which is the presupposition for all existents."

"It is like a man who finds himself on a high peak (let it be assumed that it is a rock, slippery and precipitous, rising stark (red) from a great depth to an immense height, and from the top of the peak juts forth a promontory, looking down on an unfathomable abyss). Somewhat like the experience of this man, who, moving along the edge of this rocky precipice, suddenly no longer finds a foothold or something to cling to by the hand, is the experience of the soul, it seems to me, which leaving the terra firma of diastematic notions sets out in pursuit of the pre-temporal and unextended nature. Having nothing to cling to by the hand, neither place, nor time, nor measure, nor anything else of that kind which could be a foothold for the understanding, but slipping away from every side of the incomprehensible into dizziness (*hilligia*) and helpless preplexity (*amēchanei*,) and turns back again towards the congenial, loving to know only this about the transcendent as to be persuaded that it is something other than those that are known. So when language (reason, discourse—*logos*) arrives at that which is beyond language, it is time to be silent (Ecclesiastes 3:7), and to marvel at the wonder of this ineffable power, uninterpreted and fordidden to the understanding, realizing that it was only about the works of God and not of God Himself that even the great ones (Prophets) spoke: 'Who shall declare the powers of the Lord?' (Ps 105:2), and 'I will narrate all thy works' (Ps 9:2 or Ps 117-19?) and 'Generations and generations shall praise thy works' (Ps 144:4). Of these works they speak and of these they relate the details, for declaring events which have happened they lend their voices. But when the discourse comes to that which concerns Himself who is above all conception, utter silence is what they prescribe. For they say: 'For the majesty of the glory of His holiness there is no limit' (Ps 144 : 1-5). 'Ah ! how marvellous! How the discourse fears to approach the vicinity of the knowledge of God's nature! So much so that, it does not seek to comprehend even some of the external phenomena that we can apprehend about God.' For the text does not say 'The ousia of God has no bounds', judging it too presumptuous to take even such a statement for a concept (about the *ousia* of God), but devotes the discourse merely to marvel at the magnificence in the glory that is seen around God. Even there, the discourse takes pleasure, to see not even the glory of the *ousia* itself, for even from apprehending the *holiness* of that glory it is astounded. He not only totally abandons every attempt to circumscribe the Divine nature and to say what it is, but he even does not presume to express wonder at the last of its manifestations. For the writer marvels neither at the holiness of God, nor the glory of the holiness, but only at the *majesty* of the holiness of that glory, and even there wonder seizes him. For the understanding could not grasp even that admired majesty, for it has no limit! So he says—'For the majesty of the holiness of his glory there is no limit.' So in discoursing about God, whenever the enquiry turns to the *ousia*, that is 'a time for silence' (Eccles. 3 : 7), but whenever it concerns any of the operations of the good, the knowledge of which comes down even to us, then it is the time to speak, to use words, to speak of the powers, to declare the marvels, to narrate the works, but in the matters that go beyond, it is not permitted to the creature to go outside its own limits, but it should be content if it can know itself. For, in my opinion, even the creation itself I have not known, I have not comprehended the *ousia* of the soul, what the nature of the body is, whence the existents, how generation takes place by mutual interaction, how that which is not can come to be, how that which is passes into non-being, how contradictories are fitted together in harmony in our universe. If then I do not know even the creation, how shall I discourse on that which is beyond the creation itself ? When we come to that, it is the 'time for silence'. For in these matters, silence is definitely superior. It is 'time to speak' then how our lives can be led to virtue, in Jesus Christ our Lord, to whom be glory and power, for ever and ever. Amen."[55]

This magniloquent sermon is not a simple rhetorical flourish. It

55. *In Ecclesiasten.* VII. GNO V. 411/8 to 416/10. PG 45. 729 A to 732 D. Eng. Tr. author's.
See Musurillo, *Glory to Glory*, pp. 126-129.

comes out of the depths of Gregory's own intellectual and spiritual experience. For a man gifted with a brilliant and fertile mind, the experience of seeking to comprehend the being of God must ultimately lead to (*hilligia*) or dizziness. This is not simply a borrowing from Plato who does speak of *hilligia* in exactly the opposite sense.[56] In Plato it is the multiplicity of sensible objects that makes the soul dizzy and drunk. In Gregory, it is the total absence of the sensible that leaves the soul without foothold, dizzy, eager to set foot on more familiar ground.

For *diastéma* makes it impossible for us, or even for the angels, to have a direct conception or apprehension of the being of God. Like two nights enclosing the day, the *diastéma* encloses created beings before and behind.[57] Time-space encircles the creation as a boundary which cannot be passed from inside. We can follow the concatenation (*akolouthia*) of the cosmos only up to that boundary. There the mind must stop, for the Uncreate Nature is beyond all diastematic conception,[58] has no beginning or end, and is not approachable from within the creation, either for the theological speculator or the mystic seeker.

The mind does seek to go beyond. It cannot rest within the enclosure of space-time. It knows that it is bounded and is restless in the prison of time-space. But it cannot manage to penetrate beyond by its own effort. It can guess that there is *One Who Truly Is* from whom that which exists derives its being and on whom it depends for its continuing its existence as well as for its achieving its destiny. But as Gregory says in his *Commentary on the Song of Songs*, even St. Paul, who was caught up into the "third heaven" was unable to describe accurately what he saw.

> "For all these ascents yield no clear and distinct perception and understanding of the truth (*ou theōrian te kai katalempsin enargē tēs alētheias*), but the mere hearing of the voice of the Beloved, as the Scripture calls it, the experience being characterized as 'hearing', not as knowledge by comprehension, but

56. *Hilligia* is a Platonic term. See Phaedo. 79 C. Roger Leys discusses this at length in his *L'Image de Dieu*, pp. 29-30 note. His judgment: Gregory's gnoseology is certainly not Platonic.

57. See *Contra Eunomium*, Lib. III. Tom. VII: 50 GNO II. 232/16ff. PG 45. 820 B. NPNF V.

> as something gladdening the heart. If the Bride who was elevated this much, as for example we learn about the great Paul that he was caught up into the Third heaven, is in no wise enabled to comprehend the Beloved with precision, what are we likely to experience, or in what stage can we be reckoned to be, who have not even been able to come near the outside gates of the inaccessible knowledge of God!"[59]

It would not be right to accuse Gregory of having a low view of the capacities of the human mind. It is a fact of our experience even today, after "1600 years of progress beyond Gregory", that we cannot answer the questions which he said were unanswerable in words. He gives several such lists, but we reproduce one of them below :

> "Whatever notions are hidden, such as to investigate : what is the ousia of God? What was there before the creation? What is there outside of the phenomena? What is the necessity of what happens? and whatever else the inquisitive minds seek to answer—know how to leave these things to the Holy Spirit alone, who knows the deep things of God, as the Apostle says."[60]

The recognition of the limits of human reason is no demeaning thing. It means only to be reminded that he is a creature, totally dependent upon the will of the Creator. But Gregory does not devalue reason. He grants to it enormous capacities of knowledge and skill within the created world. Unless, however, we recognize what belongs to the essence of the Christian faith, namely that God's *ousia* cannot be the object of our thought; but can only be adored and worshipped as the perfection of all Being and all Good, Man does not become himself.

The gap between creation and Creator is first ontological, but

58. *Contra Eunomium*. Lib. I. 363. GNO I. 135/13ff. PG 45. 364 D.
59. *In Canticum Canticorum* V. GNO VI, p. 138/9-18. PG 44. 860 A-B.
60. *De Vita Moysis*. GNO
PG 44. 357 C. S.c.
For other lists, see e.g.
(1) GNO I. 248/4ff. PG 45. 936 A ff. NPNF V. 257-262.
(2) PG 46. 121 B ff. NPNF V. 457 B-458 B.

not spatial-temporal, and secondly epistemological. We can neither objectify God as a vis-a-vis, nor can we know him as an object. To apply the term *diastēma* also to this gap between Creator and creation is thus slightly mis-leading, for it is not the distance between two points—one at the boundary of creation and the other at the boundary of God. For God is infinite, has no boundary. The creation is finite and has a boundary. The creation cannot exist but in God, but God is not spatial, and therefore the *diastēma* between the Creator and the creation cannot itself be conceived in any spatial terms. It is a *diastēma* between the undiastatic Creator and the diastatic or extended creation. We can conceive no mental image of such a gap, but it is our experience.

The *diastēma* between the Creator and the creation in Gregory has another unique feature—namely that it is a one-way gap. From the side of God, there is no gap. All creation is immediately present to him—in all its extension of space and time. All time and all space has come to be "at once" and are together in their entirety always present to God—or "in God".

> "The unextended (*adiastatos*), non-quantitative (*aposos*), uncircumscribed (*aperigraptos*), power (i.e. God) having in itself all the ages and all that in them are, and surpassing by his eternity, the indefinite extent of the ages... etc."[61]

Gregory makes it even clearer in the *Commentary on the Inscriptions of the Psalms* (Part II, ch XIII).

> "Through these (scriptures) we learn that to God nothing is in the future or in the past, but all things are immediately present (*toē theoē oute, ti mellei, oute ti prarōichēken, all en toē enestōti ta panta estin*)."[62]

or again in *On the Making of Man* (ch XVI end):

> "To the power of God there is nothing that has passed nor anything that is yet to happen, but even that which is expected

61. *Contra Eunomium* III. VI. 68. GNO II. 210/11ff. PG 45. 796 A. NPNF V. 209 A.

62. GNO V. 140/27-29. PG 44. 569 B.

later on as well as that which is present are equally grasped by the all-embracing power."[63]

One has to struggle with this notion intellectually and spiritually in order to be impressed by it. It is a central notion in Gregory, and explains why the *diastéma* between God and the creation is a one-way affair.

> "In the divine nature, effective power is concurrent with the decreeing will, (*epi tēs theias phuseōs sundromos esti tēi boulēsei hē dunamis*) and the will becomes the measure of the power of God (*metron tēs dunameōs tou theou to thelēma ginetai*), for the will is wisdom. And the specific characteristic of wisdom is never to be ignorant of how each thing comes to be. So the effective power becomes fully harmonized with the knowledge. When God knew that it is right for beings to come to be, together with that knowledge also ran the power to execute the existence of the beings, immediately leading that which is conceived into operative reality; there is no intervening time between the knowledge and its realization (*ouden meta tēn gnōsin huphesterizousa*[64]), but simultaneously and without time-interval along with the willing is displayed also the work which is willed (*sunēmmenōs kai adiastatōs sunanadeiknutai teē boulēi kai to ergon*). In this way the will is effective power, so that whenever he wills that the universe come into being, by that same willing that which is conceived receives the opportunity to become factually exhibited. As altogether and simultaneous is the universe and its creation conceived by God—the will, the wisdom, the effective power, and the being of beings being co-terminous."[65]

In this same passage, Gregory makes it clear that the creation itself has an *arche*, a *telos* and an interval in between, only from the perspective of those beings who are within creation, but to God Himself the beginning, the end, and the interval in between are im-

63. PG 44. 185 D. NPNF V. 406 B.

64. This word is not in Lidell & Scott's Greek-English dictionary. Gregory uses it only once—a neologism concocted from *hupō* and *husterizō*=come later.

65. *In Hexaemeron. Prooemium*. PG 44. 69 A.

mediately present. Thus from God's point of view, there is no *diastēma* between Himself and the creation. *Diastēma* is experienced only from within creation, both as internal extension in space and time, and as intermediary gap between Creator and creation.[66] Taking the opening words: "In the beginning" of the book of Genesis, he develops this idea of the simultaneity of the whole creation, past, present, and future. One of the Greek versions of the Old Testament (which Origen included in his Hexapla) was the translation of Aquila, which translates, the *berēshit* of Gen 1:1 by *en kephalaiōe* rather than the *en archē* of the other Greek versions. Gregory takes this *"in principium"* of Aquila to mean that the whole heaven and earth in all space and time was simultaneously created by God in the beginning.[67]

> "Since the Prophet (Moses) makes the Book of Genesis an introduction to the knowledge of God, the purpose of Moses is to lead by the hand those who are enslaved by sense-knowledge, through the phenomena, to that which lies beyond our sense-perception or visual knowledge which is limited to the heaven and earth; which heaven and earth as that which contains all existent beings, the Word names the last of the things which can be known through sense-perception, so that by saying that the container itself came to be through God, includes in that conception all that is contained in it to the very highest, and instead of saying that God made all at once (*athroōs*—simultaneously, in one shot) all the existents, says 'in the beginning' (*en kephalaioē*), i.e. at the very beginning (*en archēi*) God created the heaven and the earth. *Arche* and *kephalaion* mean the same thing. For both words equally mean 'all at once'. *En kephalaio* indicates that the universe came to be collectively, or in sum (*to sullēbden*) while *en arché* indicates "in a moment (*to akares*)", without a gap (*adiastaton*). For the *arché* is foreign to all diastematic notion. As the point is the beginning (*arché*) of the line and the *atom* the beginning of *mass*, the instantaneous (*to akares*) is the begin-

66. It would be interesting to make a comparative study of the Gregorian concept of *diastēma* and the sophisticated Hindu concept of *maya*. There is so much that is common to the two ideas, but much that is also fundamentally different.

67. PG 44. 69 D.

> ning of the temporally extended (*tou chronikou diastēmatos*). So the instantaneous laying the foundation (*katabolē*) of the creation by the ineffable power of God is called by Moses *arché*, i.e. *kephalaion*, in which the whole universe is said to be constituted."[68]

It should be remembered that Gregory was too sophisticated a Christian thinker to conceive heaven in spatial terms. For him heaven in the Bible usually means the boundary or horizon of human knowledge, the limit of that which is open to our senses, or the existence beyond that boundary. So in saying "In the beginning God created heaven and earth" Moses is saying, according to Gregory, that all the existent beings in the whole creation of space and time, simultaneously came into being from the hand of God.

> "All the potentialities (*aphormas*), the causes (*aitias*) and the effective powers (*dunameis*) of all existents, God laid the foundation collectively and instantaneously (*sullēbdēn ho theos en akarei kateballeto*). And in the forward motion (*hormē*—impulse) of that first will are concurrent the being of all existents—heaven, ether, stars, fire, air, sea, earth, animals, plants, all of which are simultaneously in view before the eyes of God, to Him who by the Word of His power brings them to light, to Him, as the Prophecy says, 'who knows all things even before they come to be' (Dan XIII: 42)."[69]

6. **Conclusion**

It is now time to bring this discussion on the aspect of discontinuity between the Creator and the creation to a conclusion. There are two other aspects of *diastēma*, which we reserve to another chapter, when we discuss the relation between God and Man. One is the aspect of sin as it brings a separation between God and Man, and so between God and His creation. The relation between *Sin and diastēma* in Gregory's thought needs to be further clarified. The second aspect which we also reserve to a later chapter is what the death and resurrection of Christ has to do with the *diastēma* and incarnation.

68. *In Hexaemeron.* PG 44. 69 D—72 A.
69. *Ibid.* PG 44. 72 B.

Here we have concentrated on the general aspects of *diastēma* as inherent in the very nature of the creation. What we have said so far amounts to this.

(1) All creation shares equally in *diastēma*. No created being exists without *diastēma*.

(2) *Diastēma* is totally absent in the Creator, and this is one way in which we understand his total transcendence and incomprehensibility.

(3) There is a one way *diastēma* between the creation and the Creator. God is beyond the reach of Man, both ontologically and epistemologically. The *diastēma* is a kind of ontological-epistemological envelope in which the whole creation is encased, and neither men nor angels can step out of it.

(4) *Diastēma* implies for all created beings continuous change. Change is the essential characteristic of created existence. The movement from *arché* to *telos* is mandatory for the whole creation. How this *diastēma* affects non-corporeal beings is hard for us to conceive, but it is affirmed as a matter of faith.

After having thus affirmed total discontinuity (at least one-way discontinuity) between the creation and the Creator, Gregory does not leave us there. He is also the Christian Philosopher of the Incarnation, which is the ultimate principle of continuity between the creation and the Creator. This we seek to do in the next chapter—on Participation in God.

CHAPTER V

GOD AND HIS CREATION-II
Metousia - Continuity

It is a tribute to the dialectic nature of Gregory's thought that he balances his exalted notion of the absolute transcendence of the Being of God with a daring understanding of the infinite possibility for the creation to participate in the *energies* of God. His scheme is some what as follows:

(a)	Being (*ousia*) of God	—Total transcendence—*diastēma*—separation—discontinuity
(b)	Operation (*energeia*) of God	—Immanence by will and wisdom—*metousia*—participation— continuity

While on the one hand, there is an unbridgeable ontological-epistemological gap (from our side) between the creation and the Creator, on the other hand, the creation cannot exist without participation in the will, energy, and wisdom of God. The notion of *participation* or *metousia Theou* has to be held in tension with the doctrine of *diastēma*.

The concept has been admirably studied by Fr. David L. Balás in his "*Metousia Theou, Man's Participation in God's Perfection according to St. Gregory of Nyssa*".[1]

The term *metousia*, like all Christian philosophical terms, has its antecedents in the classical culture of the Eastern Mediterranean. Everyone knows Plato's basic notion of participation: any particular existent has reality only in so far as it participates in the ideal, eternal, real, world of forms or ideas. But Plato has also an alternate

1. *Studia Anselmiana* 55. Romae 1966. 185 pp.

view. He puts it into the mouth of the Stranger (*Xenos*) in the Sophist:

> "I lay down the definition for all existents as being nothing else than power."[2]

All beings participate in the power to act and to be acted upon, according to Plato, at least in the sensible world. Participation in *dynamis* is thus the source of existence in Plato already. It is from the enquiry about the ground of the existence of all existents that Plato proceeds to his definition of existence as participation.

The job of the philosopher, particularly in the *Sophist*, is to penetrate behind appearance, through reflection, to the world of unchanging reality, where alone true knowledge is possible, and where the true light of certainty dwells.[3] For Plato this is not merely an epistemological matter, but also ontological, i.e., the Forms are the causes of particular existences, the particulars deriving their existence from the Forms, and the knowledge of the Forms is knowledge of the true reality behind each existence. But the Forms themselves participate in the reality which lies behind them—namely the good, the true Being and Cause of All. As Philip Merlan says:

> "The most striking feature of this derivation system was the derivation of physicals, i.e. sensibles, from the anterior, nonsensible, unextended, timeless spheres. As presented by Aristotle and as confirmed by what we know about Speusippus and Xenocrates, the derivation of physicals from nonphysicals was a principle accepted by Plato and his disciples."[4]

But Merlan and many other Plato scholars do not fully realize that the physicals subsist (not merely are derived from) by dependence on the non-physicals. In Plato himself this seems to have

2. *Tithemai gar horon* (*horizein*) *ta onta hōs estin ouk allo ti plēn dunamis* —Sophist 247e.

3. *Ho de ge philosophos, tei tou ontos aei dia logismōn proskeimenos ideae dia to lampron an tēs chōras oudamōs eupetes opthēnai*—Sophist 254a.

4. *From Platonism to Neoplatonism.* The Hague, Martinus Nijhoff, 3rd. ed. 1960, p. 227.

been a gradual development, less obvious in his earlier writings. The "idea" is a subjective element first, but only later on does Plato fully affirm that the forms are the "ground" of all existents. It is in Parmenides[5] that he makes his final affirmation—the Many have their being by participation in the One.[6]

With Aristotle's stringent criticism of Plato's theory of ideas, the notion of participation as the basis of reality recedes into the background, to be revived again in Middle Platonism which so profoundly influenced the Christian Fathers. The culmination in non-Christian thought of the idea of participation can be seen in Proclus (A.D. 410-485) who lived and taught a generation or two later than Gregory. For Proclus the notion of participation becomes the principle of all levels of being. The Gods are divine by having their heads, so to speak, in the transcendent realm and participating in the supreme being (*sumphuōs metechousi tōn huperkeimenōn*).[7] The other creatures participate in the Gods and are energized by them. For all souls, whether of Gods or others, their proper job is to energize and move others who can be moved *ab extra*. But all other activities of souls are *kata methexin*, by participation. Lower souls participate in the higher souls, but not in the divine life itself, which privilege is only for the highest souls. As you come down the scale, the number of participants increase, and the degree of participation consequently decreases.[8]

The idea of participation thus seems common at least to the Platonic and Pythagorean traditions in Hellenism. Even Aristotle[9] and the Stoics[10] used the word, though with less frequency and in a different sense. Plotinus, the true Oriental in Hellenic philosophy, puts matter on a lower level. Material reality's participation in being is a highly unstable and shifting relation. The soul in order

5. *Parmenides*. 157 b ff.

6. Paul Natorp, *Platos Ideenlehre*, Leipzig 1903, pp. 215ff.

7. Proclus, *The Elements of Theology*. 201 (Ed. E.R. Dodds) 2nd ed. Oxford, Clarendon 1963, p. 176.

8. *Ibid.*, Proposition. 204, p. 178.

9. M.D. Philippe, O.P. *La participation dans la philosophie d'Aristotle* in Revue Thomist 49 (1949), pp. 254-277.

10. The Stoic notion of participation was based on the idea of the unity of the cosmos, its *sympnoia*, and its all-pervading logos, in which every being participated.

to be fully participant in being, has to be "purified" from all matter, according to Plotinus.[11]

It is important for us to understand Gregory's theory of material creation as participation in contrast with that of Plotinus. We have also to remember how influential Plotinus and Proclus have been in the monastic tradition of Christianity to appreciate the genuine Christian inspiration which made Gregory depart so radically from the neo-platonist view.

Endre von Ivanka, in his *Plato Christianus,* speaks of the ideas of Neoplatonism entering the Christian world through "Dionysius the Areopagite" and influencing all subsequent Christian thought. This is more or less a true generalisation, but what we have to keep in mind is that Gregory of Nyssa before Dionysius had already examined the thoughts of a Plotinus or a Porphyry by the touchstone of the Christian category of Trinity-Incarnation, and rejected many of these ideas, and sometimes formed new ideas directly in contradiction to Plotinus.

Von Ivanka tells us that the idea of participation—"*Alle Seinsformen und Seinsstufen als verschiedene Grade und Formen des Anteilhabens an Gott aufzufassen*"[12] is a fundamental notion in Neoplatonism. Gregory, in accepting this fundamental idea, has subtly transformed it, and provided the basis for Dionysius' thought, which is significantly different from that of Proclus, as Von Ivanka shows, in its elimination of the idea of emanation (*pro-odos*). The idea of emanation. namely that each Being from its own *energeia* generates another level of Being *inferior to it,* is very much evident in the hierarchical Trinity of Eunomius. Gregory's contribution lies in separating the idea of participation from the ideas of Hierarchy of Being and Emanation, and in relating it integrally to the doctrine of creation by will and wisdom, creation not being the activity of every level of being in the same way as it is of the Creator God. For Plotinus, participation for each level of being is possible only in the next higher, and only through the hierarchical chain with the One. Gregory's contribution lies in clarifying the Christian insight that in Christ, Man, and through Man, the whole creation, directly and without intermediaries, participates in the creative energies of God Himself. He further safeguards the Transcendence of God, which

11. *Enneads* III. 6. 5.

12. See p. 254.

remains quite problematic in Neoplatonism, by putting the idea of participation or *metousia* in dialectical tension with the idea of the gap or *diastēma*, and the distinction between the *ousia* and the *energeia* of God.

1. The Nature of the Material Creation

Plotinus raises the question already in the first tractate of the First Ennead, whether it is possible for the soul to suffer. The question is intimately related to the primary question of Indo-Hellenic mysticism. Is the individual soul identical with the world-soul ? If the former can suffer, i.e. experience "pleasure and distress, fear and courage, desire and aversion", then clearly that individual soul must be quite distinct from the All-soul, which is free from all passion or experience.

In the Third Ennead, in the sixth tractate (III. 6:6), Plotinus comes to the question: If true Being is without Body, and if Body is not in true Being, given that true Being is fullness of Being and all that is outside of it is non-being, then is the Body also non-being ? Or to put the question in Plotinus' own words :

> "*Kai pōs hē tōn sōmatōn phusis mē ousa pōs de hē hulē eph hēs tautas orē kai petrai kai pasa gē sterea kai panta antitupa kai tais plēgais biazomena ta plēttomena homologei autōn tēn ousian*"[13]

Plotinus knows very well the spirit of our modern times, which is essentially the spirit of his contemporaries, the Stoics, who affirmed the being of only those that were in body and occupied some point of time and space. He recognizes that for the common-sense point of view of Aristotle (most of the time) and the Stoics only the corporeals are real, while entities like soul and being seem remote and unreal.

Plotinus takes the opposite view—the Vedantic view :

"We are thus brought back to the nature of that underlying

---

13. Bréhier's Ed. III, p. 103, lines 1-4 (III. 6:6. 33-36) "And how, the nature of corporeals not existing, whence then this matter out of which all these have come to be—mountains and rocks and all the solid earth, all these impression-creators, all that can be struck and driven by blows—surely all these proclaim the being of corporeals?"

> matter and the things believed to be based upon it: investigation will show us that Matter has no reality and is not capable of being affected. Matter must be bodiless—for body is a later production, a compound made by matter in conjunction with some other entity. Thus it is included among uncorporeal things in the sense that body is something that is neither Real-Being nor Matter.
>
> "Matter is not soul; it is not intellect, is not life, is no ideal-principle, no reason-principle; it is no limit or bound, for it is mere indetermination; it is not a power, for what does it produce?
>
> "It lives on the farther side of all these categories and so has no title to the name of Being. It will be more plausibly called a non-being ... a phantasm unabiding and yet unable to withdraw—not even strong enough to withdraw, so utterly has it failed to accept strength from the Intellectual Principle (*nous*), so absolute its lack of all Being.
>
> "Its every utterance, therefore, is a lie, it pretends to be great and it is little, to be more and it is less, and the Existence with which it masks itself is no Existence, but a passing trick making trickery of all that seems to be present in it, phantasms within a phantasm."[14]

Plotinus' view is that matter, in so far as it has any reality at all, is like a mirror, which reflects other things without being affected itself. It is not matter that is affected by the qualities that are reflected in it. The qualities affect each other, but matter remains unaffected, as matter. Matter, according to him then, is the "absence of reality". What changes is not matter, but matter reflecting one quality to matter reflecting another quality—thus only the quality is changed, matter ever remaining matter.

For Plotinus, then matter does not participate in Being. It only seems to participate in Form, but actually its essence does not allow participation. Matter is essentially evil, and cannot participate in Being which is essentially good.[15] Plotinus claims Plato's authority for the view. This is not the place to question Plotinus' views or his

---

14. Plotinus, *Enneads*. III. 6. 7. Eng. Tr. Stephen Mackenna, pp. 208-209.

15. *Enneads* III. 6. 11. *Pōs ousa kakē ephioito an tou agathou, hōs mē metalēpsei apollumenēs ho ēn.*

justification of it by Plato's authority. What Plotinus says amounts to a major contradiction—namely that Matter, as non-Being, has the qualities of the supreme Being—*apatheia* and eternity—for it is neither capable of suffering, nor can it be destroyed or dissolved.

Plotinus rightly cites Plato as saying that "matter is the receptacle and nurse of all coming to be".[16] Plato regards matter as devoid of all specific characteristics "invisible and formless, all-embracing, possessed in a most puzzling way of intelligibility, yet very hard to grasp".[17] But Plato would not have called matter evil, nor would he have affirmed, as unequivocally as Plotinus, its non-being. The transition from Plato to Plotinus seems thus clearly an influence of East Asian thought.

It is time now to look at Gregory's view on matter, to see how radically he departs from Neoplatonism. Gregory takes account of the pagan debate in his Introduction to the *In Hexaemeron* (Commentary on Genesis 1-2).

> "So, this being so, should one still not struggle to investigate and explain things regarding matter—how it came to be and whence? For we have heard some people say: 'If God is non-material (*ahulos estin ho Theos*), whence then did matter come? How did quantity come from that which has no quantity? The visible from the unseen? How the things bound by bulk and magnitude came from the unbounded? And whatever else we see concerning matter, from where did these come, not being in the nature of matter itself? For all these various problems concerning matter, we have only one solution: that all these are in no way impossible for the power of God to bring into existence, and that that power is in no way lacking in wisdom (*asophon*). But these (God's will and his power) are always together with each other, so that they are always exhibited as one, the one being instantaneously accompanied in the effective power of these operative realities, and his operative power is perfected by that wise will."[18]

16. "*Hē de hupodochē kai tithēnē geneseōs hapasēs* cited in III. 6 : 13 from *Timaeus* 49a.

17. Plato, *Timaeus* 51a. Eng. Tr. H.D.P. Lee, *Timaeus and Critias* (Penguin) 1971, p. 69.

18. *In Hexaemeron*. PG 44. 69 B.

Matter along with its qualities has as its basis the will and the wisdom of God. Matter is not alien to God, nor is it any less being than other created beings, for they too have come from the will and the wisdom of God. Matter is God's energeia and dunamis, his energy and power. But we never see matter as matter, but always only with certain qualities. Both matter and its qualities are together from the will and the power of God. It is not the case that matter is there as a resistant sub-stratum which the Demiurge then moulds into form as in the case of Plato. Thus Gregory continues:

> "If thus the wisdom and the will of God are together and identical in this way, he cannot fail to know how matter is to be found for the creation of all things, neither is he lacking in power to bring into operation that which has been conceived. He being capable of all, together all at once by his wise and powerful will he laid the foundation of all things through which matter is constituted, for the sake of the completion of existents—the light, the heavy, the dense, the rare, the soft, the hard, the wet, the dry, the cold, the warm, colour, shape, circumscription, extension—all of these are in themselves only conceptions—empty notions (not things). But none of these listed is in itself matter, but by their concurrence with each other, becomes matter. (*ou gar ti touton eph' heautou hule estin, alla sundramonta pros allēla, hulē ginetai*)"[19]

Gregory categorically denies, on a scriptural basis, the Platonic assumption of the co-eternity of matter, or of its existence apart from the things in which it is manifested.[20] Here is a clear case where the Scripture becomes normative for him over against pagan philosophy.

But his theory of matter remains problematic, in so far as he holds that without the qualities matter is non-being, and that matter is exhibited as matter only through the congruence of the qualities. However that may be, he insists that whatever existence matter

19. *In Hexaemeron.* PG 44. 69 C.

20. *De Hominis opificio.* XXIII. XXIV. PG 44. 209 C-D, 212 D ff. NPNF V. 413 B ff.

has, it is existence given to it by the will and the power of God. So matter can certainly not be evil, but must be good, for all that comes from God is good.

The creation, thus for Gregory, is the projection of the will, wisdom and power of God. God's Being is not immanent in creation, as the Stoics held. God's Being is totally transcendent, but his *energeia*, his will, his power, is the basis of the existence of the creation, and nothing in creation can thus exist without participating in God's energies. In this sense all that exists, so long as it exists, participates in the will, the power, and the wisdom of God.

Thomas Aquinas suggested that original matter was invisible, being formless, and that the Word gave form to it.[21] It was only in the thirteenth century that *creatio ex nihilo* becomes "dogma" in the western Church (fourth Lateran Council, 1215 A.D.). For Gregory in the 4th century, however, *creatio ex nihilo* is the very basis of his understanding of creation, matter, and human existence.

2. Participation in Being

Gregory makes the fundamental distinction between *ho ontōs ōn* and *ta onta*. We have no way of literally translating these into English. Between "the one who has true Being" and "those that merely exist" is the contrast Gregory wants to express by these phrases. But we need to understand this contrast more clearly—so that Being and Existence are not confused by our loose use of language.

In fact Gregory has three categories:

(a) *ho ontōs ōn* (b) *ta onta* and (c) *to mē on* being, existence, and non-being.

Of these only the first, namely God, has true being. He alone has true being in itself (*to on, ho tēi heautou phusei to einai echei*).[22] The opposite of God, Non-Being, has its existence only in appearing to exist (*to mē on, ho en tōe dokein einai monon estin*),[22] that existence being without a real hypostasis (*anupostaton echon eph' heautou tēn*

21. . . . cum fides sit de invisibilibus, etiam saecula facta sunt de invisibilibus, scilicet de materia prima, quae nuda et privata omni forma invisibilis est, et omni specie et dispositione, carens.

cited by Philip E. Hughes, *The Doctrine of Creation in Hebrews 11 : 3* in Biblical Theology Bulletin, Rome. Vol. II No. 1, Feb. 1972, pp. 64-77.

22. *De Vita Moysis*. II. GNO VII. I, 40/10-12. PG 44. 333.

phusin).[23] In between the two are those things, i.e. the universe, which has been brought from non-being into existence by the true Being, and which continues in existence, both depending on the true Being and participating in the energies of that true Being.

But this participation of the existents in the energies of true Being demands from the existent that it sheds its illusions about other existents, sees that they have no true being, disabuses itself of all non-being which falsely appears to be being, and turns to the vision and love of Him alone who has true being.[24]

On this conception of "He who Truly Is", Gregory stands squarely in the Jewish-Christian tradition and departs significantly from Plato. For Plato "the True Being" is the World of Ideas, not God. Gregory, following the Jewish-Christian Tradition and Philo in particular, gives that title only to God. Gregory has no "world of ideas" as in Plato. Even if he grants a sphere of intelligible beings, they do not have true being, but only the intermediary being between true Being and Non-Being, that is, a dependent existence.

For Philo also, who while in so many respects a true Platonist, significantly departed from Platonism in denying true being to the *kosmos noētos,*[25] it was the Revelation to Moses of Yahveh as "I am who I am" that was finally normative. The terminology that Gregory uses is Platonic, but the terms are used in the sense in which Philo had transformed them. The reality that Plato had ascribed to the "World of Forms" Philo and Gregory attribute only to Him Who is, i.e., Yahveh, who alone is *ho ontōs ōn*. Both Plato's world of forms and his world of particulars are now classed together as *ta onta*, which have come into being out of *ta mē onta*.

What then is it that holds together the cosmos and its creator? The Principle of Discontinuity we have discussed—the diastéma.

23. *Ibid,*, 23. PG 44. 333 B.

24. *Ibid.*, 26. PG 44. 333 C.

25. Philo was able to do away with the idea of *nous* which in Neoplatonism was both one and many, as intermediary between the One God and the many creatures. Prof. Wolfson shows that even the Logos, which is only "a pattern" in Philo, is not necessary as an intermediary *kosmos noētos*. All except Yahveh are created—for Philo. But even in Philo, the Platonic world of ideas have some uncreated eternity, having always existed in the mind of God. (See Wolfson's *Philo* I, p. 290). Gregory is much more strict in his affirmation of the non-eternity of creation.

The Principle of Continuity, without which the creation could not exist, is the *continuity* between the *ousia* and the *energeia* of God. Only in the light of an adequate understanding of this continuity between God and His creation can we go on to discuss further the nature of Being and the good as participation.

3. Ousia and Energeia

The heart of the controversy with Eunomius can hardly be understood without reference to the three key terms in the debate—*ousia, energeia, erga.* There is a fourth key term *dynamis*, but let us first see what Eunomius has to say about the three first. The strict logic of Eunomius could be summarized thus.

The way to measure the rank of various beings (*ousiai*) is to measure the operating power (*energeia*) of each. The measure of this power is most clearly seen in the *works* (*erga*) resulting from that operative power or energy.

In order to understand Eunomius' teaching on the relation between *ousia, energeia* and *erga* clearly, we should keep in mind that we are here speaking of the Three Higher Beings in Eunomius' Triadology—quite completely different from the Christian teaching about the Holy Trinity. For Eunomius, there are three supreme beings (not three Gods, but neither are the three in any sense one), hierarchically ordered in rank and power. Of these three only the supreme-most One is truly God. Let us put it in Eunomius' own words:

> "The whole summary of our teachings (*kath' hēmos dogmatōn*) is this: From the highest and chiefest Being is the second Being, having its being from the First, but after it, though before everything else, and a third Being, in no wise to be placed in the same class (*suntattomenēs*) as the first two, but to be regarded as subordinate (*hupotattomenēs*) to the one as to its cause, and to the other as to the *energeia* or operative power through which it came to be, it being necessary to comprehend at the same time, for the completion of the whole account, the *energeiai* which follow (*parepomenōn*) the beings (*ousiai*), and the names adhering to these. Again, because of each of these beings being absolutely simple (*haplēs*) and in every way one in being and conception, each according

to its own worthiness, circumscribed by the works (*erga*) of its own operative power (*energeia*), its effected works being the measure of its operative power, it necessarily follows that, surely, the *energeiai* which follow the various *ousiai* are greater or lesser, and also that the *ousiai* are first and the *energeiai* second in order; in sum, one can say that their difference amounts to that existing between their effected works (*erga*). For it would not be justifiable to say that it was the same *energeia* by which was produced the angels, or the stars and heaven, or again man, rather just as among the effected works some works are superior to or more honourable than others, to the same extent, it must be clear to anyone with a pious understanding, that the *energeiai* also vary accordingly for the same *energeiai* effect the same *erga*, and differences in the effected works (*erga*) manifest differences in the operative power (*energeia*)".[26]

We need not stop here to ponder over the fact that Eunomius' Trinity so curiously resembles that of Plotinus and not the Trinity as understood in the Christian tradition. Our point here is to understand the relation between *ousia* and *energeia*, Eunomius' argument is clearly that we can have direct access only to the *erga* or effected works of the three Persons, and from the measure of the *erga* we can compare the respective *energeiai* which produced them, and from the difference of the *energeiai* we can classify the various *ousiai* to which the *energeiai* are attached.

But what in this case is meant by *ousia* and *energeia*? Fortunately for us, Gregory gives a more or less precise definition of the two terms, and refers to their mutual relation.

By *ousia* Gregory means the subject of whom something is predicated, not the qualities, not the operations, but the subject of both.[27] This does not mean of course that a being is composed of a substratum or substantia to which the qualities adhere and from which the

26. *Cōntra Eunomium* I. 151. GNO I, p. 71, 28 to 73, 3. PG 45. 297. NPNF V. 50. The quotation must be directly from Eunomius' reply (apologia apologiae) to Basil's *Anatreptikos* (*Anatreptikos tou apologetikou tou dusebous Eunomiou*) against Eunomius' first apology.

27. One should be extremely careful not to assimilate the Greek notion of *ousia* to the Latin notion of *substantia*. The nuances of difference are quite subtle, the two terms come from two different ways of looking at things.

operations proceed. Without the qualities and the power of operations, the *ousia* would no longer be the *ousia* that it is. *Ousia* is not a residuum that remains when the qualities and operations are abstracted from it. The qualities and operations belong integrally to the *ousia,* but the *ousia* is more than just the congruence of the qualities and the operations. At the same time when one thinks of the *ousia*, one is not thinking of the qualities and operations, but about the Being-in-itself. Only in this sense can we understand Gregory's words :

> "Since he (Eunomius) wishes to appear very sophisticated in such matters, and spits upon those who write without handling such matters with proper logical skill, let him tell despicable little me with what logical sophistication he can know greater and lesser in pure being. What is the logic in setting forth such differences that one *ousia* is greater than another *ousia*, taking being in its proper sense (*kat' auto legō to sēmainomenon tēs ousias*)? Let him not advance any difference of qualities (*tōn poiotetōn*) or of properties (*tōn idiōmatōn*) which are comprehended around the conception behind the word *ousia*, but are alongside the subject (*para to hupokeimenon*). For it is neither the difference of vapours and colours, of weight or force, of worth or status or habits nor of anything else that we conceive in relation to bodies and souls that is the matter in hand for our investigation, but the subject (*hupokeimenon*) itself, to which the name *ousia* chiefly refers, and it is in regard to the subject that we have to see how one *ousia* is more than the other. But we have not yet heard from him, how, of the two existences that are acknowledged to *be*, so long as both *are*, how one has more *being* than the other. For each of them equally *is*, insofar as it *is*, all considerations of superior and inferior, as we have stated, being excluded from consideration."[28]

The argument is, of course, quite sophistic. But the concept of Being certainly does not coincide, in Gregory's thought, with the Latin notion of *substantia*. *Ousia* is being as such, not the substance

---

28. *Contra Eunomium* I, 180-182. GNO I. 79, 28-80, 17. PG 45. 305. NPNF V. 52 B.

of an existent without its qualities and properties, but the fact of its "*is-ness*". This is not to be confused with even the subject of which the predicates are made or from which the actions proceed.

There is a logical problem here which Gregory does not touch upon at this point. If being is "*is-ness*", and all being simply is, then how can Gregory later draw the distinction between the being of the Creator, He who is (*ho ontōs ōn*), and that of the creation, those which are (*ta onta*)? They both simply *are*, and in so far as they *are*, they are both being, and it is difficult to see in what sense the *ousia* of the Creator is superior to that of the creation. We have dealt with this problem in the previous chapter, but the logical difficulty about the conception of being remains. The fact of the matter is that Eunomius makes the same distinction between the Father on the one hand, and the Son and the Spirit on the other, as Gregory makes between the Creator and the creation. Logically it is difficult to see how the one is tenable while the other is not. In terms of the teaching of the Church, one can see the difference and justify it, but only in terms of the teaching of the Church and not in terms of pure logic. Gregory is actually using the techniques of the Second Sophistic to refute what he knows, not by logic but by Church teaching, to be false. For a Jew or a pagan, the argument of Eunomius must seem to be more convincing, and less riddled with logical difficulties. We must later examine to what extent the doctrine of "participation" overcomes some of this logical difficulty.

Let us come back to the notions of *ousia* and *energeia* and their inter-relationship, which is the matter in hand. It is best to cite Gregory's argument verbatim and then to consider it in its details:

> "It is worthwhile now, for a moment to consider how the *energeiai* 'follow' (*epontai*) the *ousiai*, what these *energeiai* are in their own nature, whether they are something else (*allo ti*) beside the *ousiai* which they follow, or part (*meros*) of them, or of the same nature as they (*tes autēs phuseōs*)—
>
> And if they are something else (*allo*), how could they have come to be alongside something (*para tinos*)? if they are the same (*to auto*), how they got cut off (*apotemnomenoi*) and instead of co-existing with them (*sunuparchein autais*) follow them externally? For we cannot learn this directly from his words

> to what extent by some natural necessity and not by deliberate choice the *energeia,* whatever it may be, has to follow the *ousia* compulsively, as heat and vapour follow fire, and their exhalations in turn are followed by the bodies which are generated from them. But it seems to me that this is not what he says—that the *ousia* of God is to be regarded as something complex and composite, having the *energeia* inseparable from and co-comprehended with itself, as the 'accident' is attached to the 'subject'; he seems rather to say that the *ousiai*, by deliberate and self-propelled movement, works out these *energeiai* from themselves. And how then does he say that that which comes to be by deliberate purpose 'follows' the *ousia* as if something externally following it? For we have not known, either in common speach or in trained discourse, such an expression—as to say that the energy of a labourer is 'following' him. It is not possible thus to bifurcate the one from the other, so that one can be conceived as existing in itself without the other; on the contrary when we say energy we co-comprehend the source which activates it; mentioning the energizer, we imply the energy also which remains unmentioned."[29]

Eunomius seems to have made a clean separation between *ousia* and *energeia* in such a way that the two become distinct entities. For him only the *ousia* of the Father is unbegotten (*agennētos*) while his *energeia* having been generated by the *ousia* cannot be regarded as unbegotten and the *energeia* is therefore not God, but something lower in rank, subordinate, to God. Gregory argues how Eunomius sees the Son as a product of the *energeia*, the latter acting somehow as an intermediary being between the *ousia* of the Father and that of the Son, but identical with neither the one nor the other. Gregory rightly points out that Eunomius' Trinity has five beings. The Fathers *ousia,* His *energeia*, the Son's *ousia*, the Son's *energeia*, and the Holy Spirit.[30]

It is at this point that Gregory makes the distinction clear between Being-from-itself, and Being-by participation. In an eloqu-

29. *CE.* I. 207-209. GNO I. 87/3 to 88/3. PG 45. 313. NPNF V. 54 B-55 A.

30. *CE.* I. 249. GNO I. 99/20-25. PG 45. 328 A. NPNF V. 58 B.

ent passage that can compare with the best in any Christian writing, Gregory sets forth his conception of the basic distinction between created Being and the Creator's Being. These are both Being only in the sense that the same word is applied to both. But the difference is indeed immense as the previous chapter has already shown. What we need to emphasize here is the basic difference in quality of Being or more accurately "mode" of Being between the Creator and the creation.

The Father, the Son, and the Holy Spirit all belong to the uncreated. Nothing else does. Only the Three-in-One have being in themselves, all others have being-by-participation. Being-in-itself is divine being, the *ousia* of the Father, Son and Holy Spirit. It is good, it is infinite; it is changeless, it cannot change towards the better, because it is perfection and there is nothing better; it cannot change towards the worse, because infinity cannot be reduced by diminution, and the lack of good cannot have any place in infinity.[31] It is not dependent on anything outside for its being or its goodness. It is good-in-itself, being-in-itself. In that it is infinite, simple and self-contained, there are no ranks of older-younger, larger-smaller, superior-inferior. It is the source of its own being and goodness, and the source of the being and goodness of all that is created. It has not come to be, it will not cease to be.

The creation on the other hand, as we have said many times, before, has both come to be from non-being. and is maintained in being, only by the will of the Creator. The participation in that will is the ground of its being. In itself it is nothing, it shares in being and goodness by participation in the constituting will which brought it out of non-being and leads it to being. It can never be infinitely good, but it can participate infinitely in the good, not because it could ever become infinite, but because the good in which it participates, which is the will or *energeia* of God, is infinite. There are degrees of greater and lesser, upper and lower, first and last in the creation precisely because it is finite. Especially in the intelligible world, the degree or rank of a being is dependent on the degree of participation in the good. This degree of participation again is determined by the degree of freedom in the will which each being possesses. Here are the words of Gregory:

31. On the nature of God's infinity see Ekkehard Muhlenberg, *Die Unendlichkeit Gottes bei Gregor von Nyssa.* Gottingen 1966, especially pp. 111-142.

> "Since the source (*pēgē*), principle (*archē*) and treasury (*chorēgia*) of all good is understood to be in the uncreated nature, the whole creation is inclined towards it, bound to it and sharing in it through communion in the First Good of the Exalted Nature (*dia tēs koinōinias tou prōton agathou tēs hupsēlēs phuseōs ephaptomenē te kai metechousa*), it follows that, the greater and the lesser in creation can be distinguished in proportion to the degree of participation in the highest, some greater, some less, which again is proportional to the participation in the autonomy of the will and its impulsion to the good. (*kata tēn analogian tēs tōn hupselōn metousias tōn men pleionōs tōn de elattonōs kata to autexousion tēs prohaireseōs metalambanontōn, to pleon kai to hētton en tēi ktisei gnōrizetai analogōs tēs hekastou hormēs*). The natures known (intelligible) in creation standing at the borderland between good and its opposite, so as to be capable of either, in accordance with the inclination of the mind, as we have been taught by the Scripture, are spoken of as greater and lesser in accordance with their dissociation from evil things and their approach towards the good, and thus having place in the higher spheres in accordance with the life of virtue."[32]

The basic and fundamental distinction thus is between good-in-itself, being-in-itself on the one hand, and good-by-participation, being-by-participation on the other. This passage of Gregory can be fatally mis-understood, if one were to go back and seek all the parallels in pagan or Jewish thought. There is no need to doubt that Plotinus' Enneads are open before him (VIth Book—at ninth tractate?). The terminology is Plotinian or Middle Platonic, but the ideas far from so. Plotinus' second Being, the Nous, is more like Eunomius' Second Person of the Trinity and is neither simple, nor being-in-itself, nor good-in-itself. For Eunomius the Son and the Holy Spirit have neither being nor goodness in themselves. Only the Unbegotten is good-in-itself, being-in-itself. The Son and the Holy Spirit have only participatory being and goodness.

The important point here for Gregory is to contradict Eunomius' argument that beings can be measured by their energies, and the

32. *CE.* I. 274-275. GNO I. 106/16 to 107/4. PG 45. 333 D. NPNF V. 60 B.

energies by their achieved results. The wind may have created a sand-dune, but the sand-dune does not explain the nature of the wind. One cannot understand the *energia* from the *erga*, for it could be the same man[33] who builds a ship and a house, but neither of these are adequate to reveal the *energeia* or the *ousia* of man. Besides, says Gregory, if the Spirit,[34] who according to Eunomius is created by the Son, is to be understood through his *erga* or effected works, to which work would you go to find out his *energeia* and his *ousia*? If we want to understand the "sky" what *energeia* or *erga* of the sky would we look for, in order that we may penetrate to its *ousia*?

Gregory in effect denies that the *ousia* of anything can be comprehended through its *energeia*. But in the case of God, it is only the *energeia* that we can know. The point is discussed clearly in the 2nd book of *Contra Eunomium* (Migne's Book XII B and XIII), in the NPNF English Translation—*Answer to Eunomius' Second Book*). Eunomius' argument of course is that the *ousia* of the Father is in His ungenerateness (*agennētos*), as the *ousia* of the Son is in His having been generated. If this were so, then it is clear that the *ousia* can be known precisely through these concepts which exhaustively define them, and also that the *ousia* of the Father and that of the Son are completely different from each other. That is in fact what *anhomoianism*, the extreme form of the Arian heresy means. If the *ousia* of the Father is unbegottenness and that of the Son begottenness, then it is clear that they are in no way even similar (*homoios*) to each other, but totally dissimilar (*anhomoios*). But the teaching of the Church is that the *ousia* of the Son is the same as that of the Father (*homoousion*), in that case unbegottenness and begottenness cannot belong to the *ousia* which is common to both. In so far as we can know, through the faculty of human conception (*epinoia*), the distinction between the Father and the Son in terms of unbegottenness and begottenness, this distinction belongs not to the *ousia* which remains unknowable. That God is, we can know. Who God is in His *ousia*, we cannot know.

This unknowability of essence is no peculiar quality of God. There are things in the created order, the essence of which we do not know—space, for example, or time, or other phenomena like

33. *CE.* I. 420-421. GNO I. 149/6ff. PG 45. 381. NPNF V. 74 A.
34. *CE.* I. 425. GNO I. 150. PG 45. 384. NPNF V. 74 B.

the human mind which we know to exist, but we know not how. We can have notions about their *ousia*, but these notions may be just as false as they may be true.

The unknowability of God's *ousia* does not mean that we cannot have certain dim and hazy notions about it, as Basil taught, and as Gregory readily acknowledges:

> "Have we not said (for we make our own the words of our Teacher) that we do have a faint (*amudra*) and scant (*brachutatē*) apprehension (*antilēpsis*, not *katalēpsis*) through our reasonings, of the divine nature, this knowledge which we gather from the names which the true faith uses regarding the Divine nature is wholly sufficient for our limited capacities?"[35]

But the terms have by no means a uniform significance, and have to be understood proportionally, i.e. with the awareness that the words used belong to our apprehension of created nature and are not adequate for the knowledge of the Creator. We do speak of God's being in terms of perfection of good, eternity, power, wisdom, etc. but our mental conceptions of these notions are taken from our experience of persons or things in creation which have similar qualities. The purpose of the language of worship (not necessarily theology) is to have a means whereby we can ascribe the glory due to God, not to capture his *ousia* in our words.

> "What is strived for in the words about God, is certainly not a well-sounding and harmonious verbal euphony, but to work out a devout understanding through which to maintain a high and worthy estimate of God."[36]

or again:

> "We say that this word-spinning (*onomatopoiia*) of men in accordance with their judgments, fitting the apellation to the subject, is a quite legitimate activity; there is nothing absurd

35. *Contra Eunomium* II. 130. GNO I. 263/21-26. PG 45. 953 B. NPNF V. 263 A.

36. *Contra Eunomium* II. 136. GNO I. 265/7-10. PG 45. 956. NPNF V. 263 B.

> in it . . . for, after all God is not works, neither has He His being in sound and speech. But God is in Himself what He is ever believed to be, but He is named by those who invoke Him, the name not being the same as what He is (for that nature is ineffable), but He has names given to Him in accordance with what is believed to be His operations in relation to our life (*ex hōn energein ti peri tēn zoēn hēmōn pepisteuetai tas epōnumias echei*)."[37]

Thus the *energeia* of God is not something "following" his *ousia*, as Eunomius would have it, but it is the form in which God's *ousia*, remaining unknowable in itself, yet impinges upon our life and is experienced by us.

Anything then that we say about God, and it is legitimate to say many things about God, can only be our speaking out what we have experienced of his operations (*energeiai*) as they touch our lives. But the *energeiai* are not external to the *ousia*. Words, however, can speak only about the *energeia* and not about the *ousia* except in so far as its operations reach us.

It is perhaps useful to point to a problem in Gregory's thought in thus making a distinction between *ousia* and *energeia*. The point at issue in the debate with Eunomius is whether "*agennesia*" or unbegottenness belongs to the *ousia* of God, and therefore whether only God the Father is strictly speaking God, whether the Son and Spirit, having their beginning from the Father and not participating in *agennesia*, can therefore be God. The Cappadocians can see the logical validity of Eunomius' argument, the conclusion of which is a direct contradiction of the teaching of the Church. For Eunomius the conclusions of strict logic are finally authoritative, as they are for many rationalists. For the Cappadocians, who have a high view of logic, logic is nevertheless not finally authoritative. The teaching of the Church is that the Son and the Holy Spirit are God, and if logic contradicts that, then there is something wrong with the way the logic functions, and the teaching of the Church becomes a criterion for evaluating the validity of the logic used. So they work out an alternate logic, the keystone of which is this distinction between *ousia* and *energeia*, and the contention that that which is known appertains only to the *energeia* and not to the *ousia*.

37. *Ibid.* 149. GNO I. 268/25-29. PG 45. 960. NPNF V. 264 B-265 A.

There are problems in this kind of logic also, to which Gregory does not seem to pay sufficient attention, probably because neither his sister Macrina nor his brother Basil were around to criticize his thought sympathetically. Is it not the *ousia* that is known through the *energeia*? In so far as we know anything at all, it is through its operations. Do we not know "corn", to take an example used by both Basil and Gregory,[38] when we know how it functions in various ways, as "seed, fruit, food and the like"? The corn is one thing, which operates in various ways. By knowing the various ways in which corn functions, do we not know corn itself? Does what we kuow about corn through the operations of corn help us to know the *ousia* of corn?

One can see that there is a resistant residuum in corn which cannot be caught in our *epinoia* about it, but can we say that because we know only about the operations of corn—i.e. its physical-chemical structure, the way it functions as "seed, fruit, food and the like"—we know nothing of its *ousia*? Can we legitimately say about God's *ousia*, that we know nothing at all about it, when so many of the operations of that *ousia* come down to us? We can very well see that God cannot be classified, that His physical-chemical structure is totally unconceivable, that He cannot be located in time or space. But when we have some knowledge of his operations, however limited and partial that knowledge may be, in what sense at all would it be correct to say that we know nothing at all of His *ousia*? This is the question to which Gregory does not appear to have given adequate consideration. Gregory does seem to affirm categorically that, at least in the case of the Lord, that the various names formed by the human mind on the basis of experienced *enērgeiai* of God, are not indicative of His essence:

> "*entautha toinun tēs zēteseōs ousēs ei ta poikilōs epilegomena tōe kurioe kat' epinoian legetai kai ouchi tēs phuseōs tēn endeixin echei.*"[39]

38. *Contra Eunomium*. II. 352. GNO I. 329/3ff. PG 45. 1028. NPNF V. 285 B.

39. *Contra Eunomium* II. 358. GNO I. 330/23-26. PG 45. 1029 D. NPNF V. 286 A.
"This being then the object of our enquiry, whether the various attributes ascribed to the Lord are said according to human conception and do not give indication of the nature."

The second difficulty in Gregory's thought is in the assertion that the Father's *ousia* is indeed unbegotten,[40] but that the unbegottenness does not belong to the essence of God, for if it did, then the *ousiai* of the Son and the Spirit, lacking this aspect of Divine nature, would be imperfect in so far as there is a lack of unbegottenness. The difficulty here lies in this: perfection of goodness, removal from all evil, indestructibility and infallibility, wisdom, power, and love—all these belong to the *ousia* of God and are shared alike by the Three Persons. All the Three Persons are *autozooi, autagathoi*, and this belongs to the *ousia* of Divinity. But if the Son and the Holy Spirit are not from themselves as the source of their *ousia* but have it from the Father by begottenness or by procession, then in what sense can they be said to have Being in themselves? Is not Eunomius' position more logical, namely that only the Father has being in himself, being uncaused, ungenerate? Everything and everyone else, including the Son and the Holy Spirit have their being from the Father. That is an absolute distinction which both sides accept. But Eunomius would say that the quality of the Father as uncaused cause is the definitive, unique, distinctive element which makes Him alone God in a manner which cannot be shared by anyone else, including the Son and the Holy Spirit.

The implications of accepting the logic of Eunomius' argument are also clear—namely to make the Nicean faith impossible, i.e. the Trinitarian belief impossible, the faith in the Incarnation of God the Son impossible, the belief in the Holy Spirit as God impossible. Here the Cappadocians make a clear choice—as St. Athanasius before them—for the faith of the Church against the demonstrations of pure logic.

And they had the intellectual equipment to seek to construct an alternate logic that would fit the faith of the Church, and the construction of that logic has played as important a role as the Niceno-Constantino-politan creed in maintaining the Faith of the Church in the Triune God and in the Son of God Incarnate.

An essential aspect of that logic was to draw the line at a different point from Eunomius. Eunomius draws his line between the *Agennetos* (ungenerate) and everything else. This is the

---

40. *agennētōs esti tou patros hē ousia.*
CE II. 379, GNO I. 337/3. PG 45. 1037. NPNF V. 288 B.

traditional place of drawing the line both in Judaism and in Neoplatonism. The line is between the One and everything else including the *nous*, both in Plotinus and in Eunomius, as well as in Philo. The Cappadocians draw the line between the Creator and all the rest. But the Creator now becomes not One in a simple way, but as Triune God, Three Persons having one *ousia*, distinguished as Three distinct *hypostaseis*, distinguished in terms of the Father in his *Unbegottenness*, the Son in his *Begottenness*, and the Spirit in his *Procession*. Thus by taking what Eunomius regarded as the essence (or *ousia*) of God, namely his unbegottenness, into the realm of *hypostasis* (note that Gregory does not say that *agennesia* is simply an operation or *energeia* of God), they create a totally new logic.

In this new logic, the distinction between *ousia* and *energeia* plays a key role, as much as the distinction between Creator and creation which is expressed by the notion of *diastēma*. The *ousia* is shared only within the godhead of the Triune Creator, and cannot be shared by the creation, either epistemologically (intellectually) or ontologically (i.e. in terms of being). But the Creator's *ousia* does share its *energeia* with the creation. In fact it is from this *energeia* that the creation has the basis of its existence and its possibility of sharing in the good. The *ousia* of God is fullness of being and infinity of goodness. By sharing in the *energeia* of that *ousia*, the creation is enabled to participate, in measured degree and not in achieved perfection, in both being and goodness.

The idea of participation has thus two levels, and is the principle of continuity alongside the *diastēma* or discontinuity between the two levels. Father Balas has given us an admirably clear account of the difference between these two levels of participation, to which we turn in the next section.

Here we need only to point out the fact that the relation between the *ousia* and the *energeia* of God remains insufficiently explained in Gregory. That the two are not discontinuous, the one being external to the other, he clearly affirms, in response to Eunomius who made *ousia* and *energeia* two distinct, almost disjunct, realities. And for Eunomius, the Son is not from the Father's *ousia*, but only from His *energeia*. The Son, for Eunomius, is unique indeed, for He is the sole product of the Father's *energeia*, everything else being the product of the Son's *energeia* and not directly of the Father's *energeia*. Here is the quotation from

Eunomius:

> "It is reasonable for one to say then that it is only to that supreme (*kuriotatē*), first (*prōtē*) and unique (*monē*) *ousi* a which received its *hypostasis* from the *energeia* of the Father, can be attributed the apellations of that which came to be (*gennēma*), that which was made (*poiēma*), that which was created (*ktisma*)... Only the Son, who was constituted (*sustas*) by the *energeia* of the Father, has his nature unshared (*akoinōnēton*) (with other beings), and is unique in his relation to Him who begat Him."[41]

The issue of the relation between *ousia* and *energeia* can be resolved only in terms of the comparison and contrast, in the Nicene faith of the Cappadocians, between the manner in which the Triune God has internal participation in the *ousia* itself, and that between the Creator and the creation where the participation is one-way and only in the *energeia.*

4. **Koinonia and Metousia**

The Son's *ousia* does not participate in the *ousia* of the Father, but only in his *energeia*, according to Eunomius, though only the Son participates in the *energeia* of the Father, while all else participate only in the *energeia* of the Son. For Eunomius both the Father and the Son are unique, the Father as sole ungenerate, and the Son as sole generate from the *energeia* of the Father, but their respective *ousiai* are unshared by the other—*akoinonēta.* The Son has his being by participation (*metousia*) in the *energeia* of the Father, but not by sharing (*koinōnia*) in the Father's *ousia,* which as *ungenerate* cannot be shared at all.

This distinction between *koinonia* and *metousia* is fundamentally accepted by Gregory, but, as we have said, he draws the lines differently. The Being of God, variously expressed by the neuter *to on* or more usually by the masculine *ho ontōs ōn,* is God's *ousia,* which has no name. For Gregory, the Three Persons in the Triune God have *koinonia* in this *ousia,* but it is not shared by the creation, at least not in the same way. The creation "abides" (*menei*) in

41. *Contra Eunomium* III. II. 73. GNO II. 76/8-15. PG 45. 644. NPNF V. 160 B.

the Being of God, without which abiding it cannot have any existence: (*ou gar an ti diamenei en tōe einai, mē en tōe onti menon*[42]). The Being of God alone is "proper and primary being" (*to de kuriōs kai prōtōs on hē Theia phusis estin*), the necessary precondition for the existence of all other things (*hēn ex anagkēs pisteuein en pasin einai tois ousin hē diamonē tōn ontōn katanagkazei*).[43]

God's *ousia* does not create and hold the creation in being by imparting his own *ousia* to them, but rather by his *energeia* (operation), *thelēma* (will), *dunamis* (power), *sophia* (wisdom). In some ways *energeia* is the comprehensive term which includes all the others, and this *energeia* is not external to the *ousia*. For the creation to be in the Being of God, does not strictly mean being in the *ousia* of God, but only in the *energeia*. Here we cannot use spatial metaphors like the *energeia* being "internal" to the *ousia* of God, but then neither can we say that it is external, for to be in the *energeia* of God is to be in God.

> "It (Scripture) teaches through these (i.e. Ps 106 : 39/40) that to be in Him who is, is truly to be (*to en toē onti einai alēthōs estin einai*). If one has fallen out from Him who is, he is no longer in being (*ei ti tou ontos ekpeptōken, oude en toe einai esti*). For to be in evil is not properly to be (*to gar en kakiae einai ouk esti kuriōs einai*) for evil itself, in itself, is not (*autē kath' heautēn hē kakia ouk estin*), but the non-existence of good becomes evil. So then, just as to be in Him who is, is to be in being, so also he who becomes in that which is not (for this is evil) is abrogated from being (*exoudenōtai*)."[44]

But the "being in God" cf the creation is not of the same kind as the "being God" of the Three Persons of the Trinity. To be in the *ousia* of God is possible only for the Three Persons of the Trinity.

What Fr. Balás has clearly shown is that the idea of participation should be distinguished in its two meanings. There is participation by nature, and participation in the qualities of another

42. *Oratio Catechetica*. 32. 6. Srawley, p. 118/2-119/1. PG 45. 80 D. NPNF V. 500 A.

43. *Ibid.*

44. *In Inscrip. Psalm.* I. VIII. GNO V. 62/26-63/6. PG 44. 480 A.

nature. This Patristic distinction is the one which was later unrecognisably distorted into that false distinction between nature and grace, as if nature itself was not a gift of grace.

Here we come to the heart of this dissertation. The point is best illumined by a rigorous examination of the views of Fr. Balás on Gregory's idea of participation. His major conclusions are here reproduced in order to facilitate the examination. The theological framework from within which Fr. Balás looks at Gregory's thought is already alien to that thought, as is clear from the following passage in his chapter on "Conclusion":

> "The positive achievements of Gregory should not make us overlook some imperfections or shortcomings in his theology of participation. Although he clearly distinguished the uncreated and the created spiritual beings, and did not confuse human nature as such with the participation of the divine perfections, he yet *failed to distinguish between natural and supernatural participation* and also tended to minimise the distinction between present anticipation and the future eternal fulfilment."[45]

Fr. Balás here makes the usual distinction between natural and supernatural, the bane of western theology, and makes that a criterion for evaluating the perfections and imperfections of Gregory's thought. It is significant, however, to note first that the natural-supernatural distinction is totally alien to Gregory's thought, as we hope to make clear in the next chapter. Here it needs to be pointed out that the Greek language, which Gregory used, has no word for "supernatural". A Greek can speak about something marvellous or strange as *huperphuēs*, but he does not use an adjective like *huperphusikos* or *huperphusikē*. Gregory does use, though extremely rarely, the verb *huperphuō* (only once in the volumes published so far of the critical edition)[46] and the adverb *huperphuōs* (only twice in the published volumes of the critical text)[47], but always in a literary and not philosophical sense. One searches in vain for the concept of the "supernatural" in Gregory. From

45. *Metousia Theou*. Rome 1966, p. 165. italics ours.

46. GNO IX. 260/28 (*In Sanctum Pascha*).

47. GNO VI. 141/6 (*In Canticum Canticorum*). GNO VIII. 377/11 (*Vita Macrinae*).

Gregory's point of view, "supernatural" is an impossible concept. For anything to be supernatural is, for Gregory, disastrous and suicidal. For Gregory there are only two basic kinds of nature—the uncreate and the created. The uncreate nature cannot be "supernatural", for there is nothing that is "supra" to its nature. The created nature cannot be "supernatural", for nature is the boundary given to each being by the good pleasure of its Creator, and its being what it is depends on remaining within that nature. To go "outside" or "above" that nature is immediately to cease to exist.

This has been the major problem with Gregorian thought in the west. Loaded with categories like "original sin" and "supernatural", baggage accumulated through centuries of alienation from the authentic tradition, theologians have lost the ability to see straight and to look at the profundity of a thinker like Gregory who does not operate within that framework.

Fortunately Fr. Balás' work is not deeply marred by the remark which we have cited from his concluding chapter. Whatever may have been his intention, the paragraph may come to his rescue when and if he is accused of departing from the western tradition. His own analysis of the notion of "participation" in Gregory has been only slightly affected by his perspective, and it remains the best basis for further study of the subject, which after all is the stated intention[48] of his dissertation.

Fr. Balás' main contribution to clarifying the notion of participation is in distinguishing between "vertical" participation and horizontal participation. This idea is more clearly presented in his paper read before the American Catholic Philosophical Association from which we quote :

> "If we ask now for the role of participation in Gregory's universe, we find that it has two main directions, which we may call respectively 'vertical' and 'horizontal'.
>
> 1. 'Vertical' participation is not found, as we might have expected, between the sensible and the intellectual natures but rather between the created and the uncreated, and especially between the created intellectual and uncreated intellectual. Thus vertical participation is primarily the relation of spiritual creatures to God.

48. *op. cit.*, p. 167.

> 2. 'Horizontal' participation, on the other hand, is found on each level of reality and consists in sharing in the common nature of the species."[49]

Fr. Balás rightly remarks in the same article that Gregory's thought is "a theology of participation". And it is as such that we have sought to understand it, though we would qualify that statement by saying that Gregory's thought is a fully Christian attempt to see the whole of reality (not just theology) in terms of *diastēma* and *metousia,* or distension and participation, and not of participation alone.

What is meant by "horizontal" participation? The concept is in some ways ambiguous and inadequate. Take the nature of Man as a concrete example. It is evident that all members of the human species participate "horizontally" in some common characteristics. But what looks so evident becomes problematic when closely examined. Can we say that having two legs, two hands, two eyes, two ears, one head, one nose, and so on are part of this common participation? For we run immediately into two difficulties. First, many of these characteristics are shared by other animals than men, and are in some sense horizontal participation, not in the nature of man, but rather in the nature of animals. This means that man's participation in certain characteristics which are common to all men, also result in a "downward" participation in "lower" natures. Secondly, the absence of one or more of these external qualities does not abolish his humanity. We cannot regard a man with one leg and one eye as something less than a man.

These two difficulties together lead to the question. which is already awkward and minimalist. "What is the essential element of horizontal participation that makes a man Man ?" Once we have delineated some "essential" element, we are in danger of regarding the other elements as somehow "non-essential" or accidental. We shall look at this problem of the nature of Man more closely in the following chapter.

We need pause here only to raise the question whether the

49. *Christian Transformation of Greek Philosophy illustrated by Gregory of Nyssa's use of the Notion of Participation*, in Proceedings of the American Catholic Phllosophical Association, Washington, D.C. 1966, pp. 152-157, see esp. p. 154.

distinction between horizontal participation and vertical participation is itself brought in from the dichotomous category-structure of "nature and grace" or natural and supernatural. The vertical-horizontal terminology also is in any case external to Gregory, and it seems more useful to see the distinctions in participation by seeking them through the Gregorian category-structure of *ousia-energeia* as we shall do in the next chapter.

5. Summary

Across the *diastēma* or discontinuity between the Creator and the creation, there exists the continuity of *metousia* or participation. Without that participation, nothing can exist.[50] But participation in Being, in Life, in the good are all three inseparable from each other. Participation in the *ousia* of God means to be *autozoēs, autagathos, ho ontōs ōn*. This is possible only for the Three Persons of the Triune Godhead. What we can participate in is the being, life, and goodness of God as it is given to us in God's *energeia* which has brought us into being, sustains us in life, and leads us in the good. All three belong to the nature of man, and the whole of the nature of man is God's gracious gift. Nature is grace.

But participation in Being, Life, and the Good as the original possibility of unfallen Man, has been lost, because of Man's choice to participate in Evil, Death, and Non-being. The restoration of that possibility, as Gregory sees it, is the whole point of the Incarnation of the Lord.

50. This point is recognized by Fr. Balás, but minimised. He cites several passages in Gregory which speak about participation in being, and devotes a whole chapter to it (ch IV—*The Participation in Being*), but then in the conclusion of the chapter he makes the strange statement: "It is, however, only in a few texts that Gregory expresses this ontological dependence of all beings on God in terms of participation. In most cases 'participation of the real Being' serves to describe our (the intellectual nature's) communion with God, which is not simple existence but eternal happiness or at least its anticipation." p. 120) Here the author seems to lapse into some kind of scholasticism which seems so alien to the world of Gregory. Existence and Beatitude cannot be so easily separated. Ontology and spirituality are inseparable in Gregory.

CHAPTER VI

GOD AND MAN
Continuity and Discontinuity

The two fundamental principles we have outlined in the previous chapters are the basic terms in which Man can also be understood—though there are also many related terms which we need to examine. The *diastéma* constitutes the difference between God and Man, but equally to that difference belongs the fact that whatever Man has or is, it is only by participation in the *energeia* of God, and not out of himself.

Gregory has been accused of having an inadequate doctrine of grace. That accusation can be made only by those who have tended to regard nature as a self-sufficient and self-contained unit with its own laws, into which grace has to "intervene" from outside. One heard this charge made from a Lutheran perspective by a competent Gregory scholar at the Second International Colloquy on Gregory of Nyssa at Freckenhorst (Germany) in September 1972. Dr. Martin Ritter of the University of Gottingen, in his paper on *Die Gnadenlehre Gregors von Nyssa*, based on an analysis of Gregory's *de Vita Moysis* alone, made the true observation that Gregory's "interpretation of the essence and significance of divine grace differs profoundly from Augustine's conception." With that judgment one has to agree, and perhaps also regret the fact that Augustine did not have the benefit of fully understanding Gregory's thought, from which he seemed to have picked up several stray ideas, possibly through intermediaries like Jerome.

Ritter found the main points of difference in the absence of three elements in Gregory: (a) a distinct psychological concept of the will, conceived in accordance with the picture of the divine will, (b) the notion of original sin, and (c) pre-destination. He is certainly right about the two latter points, which make Augustine's thought-structure entirely unacceptable for the present writer. But to suggest that Gregory does not have an adequate notion of grace or that he ignores the psychology of the human will is most likely the consequence alike of one's own pre-conditioning and the faulty method of discovering Gregory's thought-structure from only one of his many works.

On the question of grace, we are in a position to document the fact that for Gregory, all is grace. There is no nature apart from or outside of grace. The relation between the grace of God and the human will shall be further discussed in this chapter, under the heading *sunergeia*. Here we need only cite a few sentences from Gregory's treatise *On the Christian Mode of Life*. We need to remember that for Gregory, if God's grace were to act in such a way that man could not resist it and that no co-operation and assent from man were necessary, then this would reduce man to the level of animals and other objects. Precisely that capacity to respond to the grace of God by free consent distinguishes man from other animals and inanimate things which are also dependent on the grace of God both for their existence and for whatever is good in them.

But Grace certainly is not automatic:

> "The grace of God does not naturally frequent souls which are fleeing from salvation, and the power of human virtue is not sufficient in itself to cause the souls not sharing in grace to ascend to the beauty of life. For it says: 'Unless the Lord build the house and keep the city, he labours in vain that builds it and watches in vain who keeps it.' (ct Ps 126:1). And again: 'For not with their own sword did they conquer the land, nor did their own arm make them victorious (although they used their swords and arms in their struggles), but it was Your right arm, and the light of Your countenance." (Ps 43:4) What does this mean? It means that the Lord from on high enters into an alliance with the doers, and at the same time, it means that it is not necessary for men considering human efforts to think that the entire crown rests upon their struggles, but it is necessary for them to refer their hopes for their goal to the will of God."[1]

1. *De Instituto Christtano*. GNO VIII. 1. 47/7-22. Eng. Tr. from Virginia Woods Callahan, *St. Gregory of Nyssa, Ascetical Works*. Washington 1967, pp. 131-132.

We begin with this reference to the question raised by Dr. Ritter about the understanding of grace in Gregory of Nyssa, for only when we are able to grasp his view of the relation between God and Man, we can begin to see how his doctrine of grace and doctrine of *sunergeia* fit into the structure of his thought. And Gregory's doctrine of grace, as we shall see, remains characteristically in dialectical tension with the notion of freedom to which we should now turn. If there is only grace (*sola gratia*) and no freedom for man, then man descends to the level of beasts and things. Freedom is the essential constituent of Man as man, and any doctrine of God that puts the sovereignty of God above freedom itself, is in some contradiction with the faith of the Church, at least for Gregory of Nyssa.

1. The Freedom of God and Man

What could we mean by saying that God is free? Since we have dealt with this question in an earlier work,[2] here we need only more precisely to document the ideas briefly expressed there.

The idea of Freedom is discussed by Gregory in all his works; we shall cite here half a dozen samples. The first occurs in the Dialogue *On the Soul and the Resurrection*, and gives a definition of freedom:

> Freedom is the ecsimilation (*exhomoiosis*) to that which has no master (*adespoton*) and is self-regulating (*autokrates*), which was God's gift to us in the beginning, but which is now obscured along with the shame of being condemned. All freedom is one in nature, homogeneous with itself. Therefore it logically follows that all that is free will be harmonized with all else that is free.[3] Virtue has no master. Therefore all that is free is in virtue, for the free has no master either.[4]

2. P. Verghese, *The Freedom of Man*, Westminster Press, Philadelphia 1972, pp. 73-103.

3. *Akolouthōs oun pan to eleutheron tōi homoiōi sunarmosthēsetai.*

4. *Aretē de adespoton. Oukoun en tautēi genēsetai pan to eleutheron, adespoton gar to eleutheron.* PG 46. 101 C-104 A. NPNF V. 452 A-B.

The second passage is in *On the Making of Man:*

> "This (to say that Man was made in accordance with the image of God) is the same as to say that He made human nature participant in all good. For if the Divine (nature) is the fullness of all good, and this (the human) is the image of that, then the image has its resemblance to the archetype in being full of all good. There is thus in us the form (*idea*) of all good, of all virtue and wisdom, and all that is conceived in the higher nature. But first among all these is necessarily Freedom (*to eleutheron*) and not to be subjugated to any power of nature (*mē hupezeuchthai tini phusikēi dunasteia*), but to have in one's own control the capacity to turn will and knowledge to whatever one chooses. For virtue is something that acknowledges no master (*adespoton*) and acts of free will (*ekousion*) since that which is compelled and forced can never be virtue."[5]

The third passage is taken from the *Great Catechism,* clearly showing how the ideas of the Freedom of God and the Freedom of Man are keys to Gregory's thought-structure, pervading all his philosophical works:

> "He who made man for participation in his own good, and built into his nature the potentialities (*aphormas*) for all good things, so that by each of these possibilities he might be impelled to the corresponding good, would never have deprived him of that most excellent and most honoured of all goods, I mean the grace of having no master and having one's authority in oneself. For if some kind of necessity overruled human life, the image would have been partially falsified in this respect, being estranged from its archetype by this dissimilarity. For how can that which is subjugated and enslaved by some necessity be named an image of the Royal nature? Thus that which is in every way made similar to the Divine nature must needs be self-ruled and without a master in its very nature, so that its participation in the good

5. *De hominis opificio*. XVI. PG 44. 184 B. NPNF V. 405 B.

> would be the prize of virtue (*hōste athlon aretēs einai tēn ton agathōn metousian*)"[6]

A fourth passage from the same *Great Catechism* raises the question of freedom in the context of the question why evil persists even after the saving economy of Jesus Christ has been effectively carried out. Gregory's answer is that if all men were to be saved in spite of themselves by some "sovereign grace" such as Augustine would have conceived, then man would by necessity become good and thereby lose his freedom. Salvation for Gregory can never be *sola gratia,* but only by the interaction of the grace of God and the freedom of Man. Gregory makes clear that the grace of God has appeared to all men, but some men in their freedom refuse to accept the grace:

> "If then impartially the call has come to all, without distinction of rank or status or age or racial or national differences (for that is the meaning of Pentecost, the Gospel being presented from the first in all the languages of those present), how can anyone still find fault with God, against all good reasoning, for the fact that the word of the Gospel has not become effective in all? For He who has sovereign authority over all, out of the great respect for man, left something to be within our own authority, something of which each of us ourself alone can be lord (*ou monos hekastos esti kurios*). This is the faculty of will or choice (*hē proairesis*), something not slavish, being its own authority (*adouloton ti chrēma, kai autexousion*), seated in the freedom of the mind (*en tēi eleutheriae tēs dianoias keimenon*)."[7]

We will cite a fifth passage which is suprising in its modernity, though the rhetoric may strike the modern ear as a bit excessive. It can be understood as a most eloquent defence of the freedom of every single human being, unparallelled in classical pagan or Christian antiquity. To see it merely as an attack on the institution of

6. *Oratio Catechetica.* 5. Srawley. 26 : 3-27/3. PG 45. 24 C-D. NPNF V. 479 B.

7. *Oratio Catechetica.* 30. Srawley, 112/1-14. PG 45. 76 D-77 A. NPNF V. 498 B.

slavery is to reduce its power to mere social criticism. The context is an interpretation of Ecclesiastes 2:7a, which speaks of the Rich Man's bragging;

> "Slaves and servants I have purchased,
> And home-born slaves also I have begotten."[8]

And this special feature of a rich household, to which Gregory of Nyssa was not a stranger, of having sumptuous furnishings and many slaves and servants (a feature of human existence that has not totally vanished from the earth, but only takes new forms in our time), becomes the occasion for a frontal attack on the very institution of slavery and on the exploitation of men by men as an affront to human freedom, as a denial of the nature of man, as a rebellion against God's order in the universe:

> "'Slaves and servants', says he, 'I have purchased and home-bred slaves also I have begotten.' Do you see the bulk of your inflated self-importance? You would lift yourself up, by these words, against God Himself! For the whole creation (*ta sumpanta*), we have heard from the prophecy, is owned by Him who transcends all. He who makes the property of God his own property, does he not transgress the limits of the power of his species, when he presumes to be owner of men as well as women? Does he not, by his arrogance, transgress nature itself, by regarding himself to be something else than those who are so ruled? 'I have purchased slaves and servants'! What are you saying? By condemning to slavery man whose nature is free and self-authoritative (*hou eleuthera hē phusis kai autexousios*), you are setting up laws contrary to God, upsetting the very law of nature. Him who was made to be lord of the earth, ordained to rule, you bring under the yoke of slavery, thus rebelling and fighting against the very order established by God! You forget the bounds of your authority, for up to the dominion over the irrational creation is the limit of your rule. 'He was given dominion', says the word, 'over birds and fishes and quadrupeds and reptiles'. How do you dare to go beyond that limit and exalt yourself

---

8. Ecclesiast 2:7 LXX

also over the free nature, making them akin to the innumerable quadrupeds and sine-peds (reptiles without legs)? Everything has been subordinated to man, shouts the word of prophecy, and includes in the number of those subordinated beasts and cattle and sheep. Have beasts given birth to human beings for you? Or have cattle borne for you human offspring? The slaves of Men are irrational beasts alone. To you, this is not enough? 'He brought forth grass for the animals and plants for serving man'. You have disrupted the natural distinction between servitude and mastery, and have made some to serve their own kind and some to be masters of their own race! 'I have bought slaves and servants'! For what price, pray, tell me. What have you found, among existent beings, as an adequate price for this human nature? How much money have you estimated as the value of a rational being? How many *obols*[9] would you regard as a fair price for the image of God? For how many *statera*[10] would you sell the God-fashioned nature? For God said 'Let us make man, according to our own image and likeness'. Him who is thus the likeness of God, having dominion over the earth, who has inherited full authority over everything that is upon the earth, who is it, tell me, who will sell him and who will buy him? Only God is able to do this. Or perhaps not even God. For 'God does not repent of his gifts', it says. Therefore he who subjected us to slavery cannot be God, for even when we by our own choice have been enslaved to sin, God called us back to freedom. If even God does not enslave the free, who is it that regards his own authority as greater than God's? For how can he who is the ruler of the whole earth and of all that is on the earth, be 'sold'? For then all the possessions of him who is sold would also have to be handed over along with him. How much shall we estimate the price of the whole earth? And how much the price of all that is on the earth? If these are invaluable, tell me, what is a fair price for buying these? And if you say, the whole cosmos, even this you will not find to be an adequate price. For even if the whole cosmos is set against the value of human nature, it is not a

9. one-sixth of a Greek drachma, a small coin.
10. a silver or gold coin.

sufficient price for the human soul. When Man himself is offered for sale, it is he who is the lord of the earth that is led to the auction-stall. Therefore along with men is offered for sale also the whole creation which is subjected to him. That means the earth and the sea and the islands, and all that are in these. Who will then hand over that which is sold? Who will then take over that which is purchased, along with all the property that is part of the transaction? Ah, but that little certificate, that written deed of sale, the counting out of a few *obols*, these have deceived you into thinking that you are the master and owner of the image of God? O, what stupidity! If you lost that little certificate, or if that deed of sale was eaten up by moths, or water coming in somehow blotted out the writing, what evidence would you have for having the slave? What would then be the proof of your ownership? For I see nothing more than the name that is added to him who is in your custody from your name. For who can it be that added to your nature more authority? Neither age nor beauty, neither fitness of body nor superiority in virtue. Of the same origin as you are, biologically the same as you: both you who lord it over him and he who is subjugated to your lordship, are equally subject to the same passions of soul and body, the same pains and pleasures, the same mirth and anguish, the same joys and sorrows, the same desires and fears, the same diseases and death."[11]

It should be clear from this passage that for Gregory man as man is free in his nature, not simply because he is a Christian, or because he has been redeemed by Christ. By his very creation he is Lord of the earth. That is his true nature, even when he is a poor slave of some rich lord. Sin is not the true nature of man, but freedom. And every man, every last man, is free.

Just one more text from Gregory, though many more could be adduced. This one is from the work called *De Mortuis*, about not weeping over those who die in the faith:

"Since man has come to be as a theomorphous being *theoeidēs*, honoured with the blessedness of being self-authoritative (for

11. *In Ecclesiasten.* Oratio IV. GNO V. 334/15-338/8. PG 44. 664 C-665 D.

the specificity of God's blessedness is in that it is self ruling and owns no master), it follows necessarily that when he is compelled by force to do something against his self-ruling will, that is a deprivation of his dignity. For if the human nature, by its own free choice, contravenes its self-ruling power and is forced by compulsion to do something that is not agreeable to it, this is tantamount to deprivation of its original goodness and a violent filching of its God-like honour (*tēs isotheou timēs aposterēsis*). For it is indeed God-like to be self-ruling (*isotheon gar esti to autexousion*). In order, therefore, on the one hand, that authority may remain in the (human) nature, and on the other that evil may be abrogated (from it), the wisdom of God has seen to it to permit the human mind (*epinoia*) to move towards that which it wills, so that having tasted the evils that it has desired, and learned from experience what it has exchanged for what, it may run backwards along its desires by its own will towards the original beatitude, casting away from its own nature, as if getting rid of a heavy burden, all that is subject to passion and brutish, from which it is cleansed and purified, either by attention to perfection and by love of wisdom in this present life, or by being purified in the fiery mould after departure from this life. (The even longer sentence which follows gives the illustration of a medical doctor who has all the knowledge and skill necessary both to heal and to harm, and has still to learn many things by trial and error, causing a lot of damage in the process of learning. So the Father has given us the knowledge of good and evil, but we have chosen to learn the hard way, because of boyish mischievousness and inexperience).[12]

In the analysis of these six passages, we notice that Man's nature cannot, for Gregory, be understood except in terms of being the visible, manifest expression (image) of God's *ousia*. Man has not only his *ousia* from the *energeia* of God, but his *ousia* is a visible reflection of the invisible *ousia* of God. What we can understand about God's *ousia* is therefore whatever we can find to be good in man. The primary quality of God's *ousia*, we understand

12. *De Mortuius*, GNO IX. I. 54/1 to 55/11. PG 46. 524 A-C.

from the Scriptures, according to Gregory, is *makariotēs*, blessedness, beatitude, pure joy, *ananda*. The meaning of the word we translate "beatitude" or "blessedness" is certainly beyond exhaustive formulation. Gregory breaks into rhapsody when he wants to say something about what beatitude means:

> "That which is truly blessed is the divine itself. For what we consider as being itself is in fact this blessedness. That undefiled life, the ineffable and inconceivable good, the indescribable beauty, grace-in-itself, and wisdom and power, the true light, the source-spring of all good, the all-transcending authority, the only desirable and loveable, that which is ever the same, unbroken joy, the eternal gladness, about whom having said all that is possible to say, one has still said nothing adequate."[13]

In this same homily on the Beatitudes Gregory goes on to say:

> "Since then, He who fashioned man, made him in accordance with the image of God, he (man) is a second blessed by participation in the Truly Blessed, having come to be in the likeness[14] of that blessedness. Just as in physical (sculptural) beauty of form, the prototype of the beauty is and exists in a living person, and this is only a duplicate form of that of which it is the image, so also human nature, being an image of the transcendent blessedness, is imprinted by the same goodness and beauty, when in itself it shows forth the characteristic qualities of that blessedness."[15]

This then, according to Gregory, is the nature of Man. Image of God means participation in the very characteristics of God—to be the perfection of all good, all beauty, all love, all wisdom, all power. This, rather than sin, as in Augustine, is the decisive and distinctive element in Man. What about sin? Does Gregory totally ignore sin? No, far from it. Sin is that which explains

13. *De Beatitudinibus* I. PG 44. 1197 A.
14. reading *omoiōmati* instead of *onomati*.
15. *De Beatitudinibus* I. PG 44. 1197 B.

the discrepancy between what Man is, as he now finds himself, and what Man is, as he ought to be, as he was created to be. Sin does not appertain to human nature, but is something which has accrued to it, something alien, which alienates, changes, falsifies human nature.

Gregory deals with this problem of the discrepancy between what man is today found to be, and what he was originally meant to be, in several different passages. We can cite only one or two here:

> "What does this word image mean, perhaps you ask. How can the corporeal be likened to the Incorporeal? How the evanescent to the Eternal? The changing and altering to the Unchanging? the passible and corruptible to the Impassible and Incorruptible? that which always lives mixed with evil and grows up with evil, to that which is untouched by evil? Great is the difference between that which is conceived in the archetype and that which has come to be in accordance with its image."[16]

or a little later:

> "How come that the Divine is in Blessedness, while the human is in misery, and yet the Scripture names the latter the likeness of the former? So we should examine the words carefully, for we shall find that that which came to be in accordance with the image is one thing, and that which is now manifested in wretchedness is another. 'God made', says it, 'man, in accordance with the image of God He made him'. That is the end of the creation in accordance with the image. Afterwards Scripture makes a resumption of the narrative of fashioning, and says: 'Male and female he created them'. I believe it is evident to all that this aspect is something outside what we know about the prototype. 'For in Christ Jesus', as the Apostle says, 'there is neither male nor female'. But the word says man is thus divided. So it seems the fashioning of our nature is somehow two-fold, one likened to the divine, the other divided according to this distinction."[17]

16. *De hominis opificio.* XVI. FG 44. 180 B. NPNF V. 404 B.
17. *De hominis opificio* XVI. PG 44. 181 A-B. NPNF V. 405 A.

Here Gregory comes to the heart of his conception of Man—that he is in every sense a middle being, torn and pulled in two different directions, under mandate to use his freedom to find his equilibrium between these two opposing forces. On the one hand, his deeper, original nature is that of being like God, full of all goodness and love and power. But there is a secondary level of his existence, in which he is divided into Man and Woman, like animals, corporeal, subject to passions, deathbound, irrational."[18]

It is quite evident to a modern student of the Scripture that Gregory's exegesis of the Genesis account is quite strained. Karl Barth[19] seems to understand it more correctly when he sees the two expressions about creation in the image and creation as male and female, simply as two different ways of saying the same thing—the phenomenon known as "Hebrew parallelism". And so purely from an exegetical perspective Barth is right in seeing the image precisely in the man-woman distinction, and in the possibility of personal mutuality and love.

The problem with the Barthian exegesis, however, is that the male-female distinction applies equally to most other species of animals, and could not have been in the mind of the narrator of the Genesis account as the distinctive element in the nature of Man. So while we must respect the validity of the claim that the man-woman relationship belongs essentially to God's act of creation, we must also acknowledge the fact that that could hardly be regarded as constituting the image.

Gregory says quite definitely that the man-woman relationship is not the result of sin, but something provided by the Creator in view of sin. We should discuss the place of the human body ("coats of skin"), death and sex later on. But here the point is simply to say that freedom belongs structurally to human existence, and Gregory sees freedom in a much larger context than Augustine, for whom only freedom from sin is really important.

For Gregory on the other hand, freedom belongs to the very *ousia* of God, and man's job in the universe is to exhibit that freedom in the human *phusis*.

In what then does God's freedom consist? In his transcendence, in his immanence by will, and in his creative power.

18. *De hominis opificio. loc. cit.*
19. See *Kirchliche Dogmatik* I.

By transcendence Gregory means that he is not so linked to any causal chain that something happening along that chain would affect him in turn. This is the true meaning of *apatheia.*

By immanence by will, Gregory means that while the cosmos does not affect the Creator in any way, the cosmos itself derives its *arché*, *telos* and the *dynamis* to move from *arché* to *telos*, from the will of God, therefore that the whole creation is present to God as an expression of His will, and that He is present in it not by his *ousia* but by His creative will. God's *ousia* is free from the creation, but His will, dynamic, free, creative, is the ground of the creation.

By creative power, Gregory means that God being the fountain and source-spring of all good, his freedom constantly brings forth new good into manifestation through the time-processs. We discuss each of these three points in some detail.

Man, in so far as he is the image of God, has to participate in all three aspects of God's freedom, though the mode of expression of man's freedom is fundamentally different, since he is this side of the *diastéma.*

2. Freedom as Transcendence

The Gregorian notion of freedom has of course to be understood in relation to the notion of Fate (*heimarmene*) which was an essential idea of Stoicism.[20] This view of a necessary causal chain connecting all things and all events in one *symploké* was reflected not only in the Stoic idea of fate or causal necessity, but also in the Gnostic idea of pre-destination which became so normative for Augustine and the Reformers of the 16th century. Origen had attacked the pre-destinarianism of the Gnostics as a violation of the freedom and dignity of man.[21] Gregory follows the same line in

20. See the Study by Dom David Amand, *Fatalisme et liberté dans l'antiquité grecque, ch. IX. Grégoire de Nysse*, Louvain, Univ. Bibliothèques, *Recueil de travaux d'histoire et de philologie* (1945) 3, 19. pp. 405-439.

21. The idea that all human beings are created equal, and that the diversity among people originates not in pre-destination, but in the free decision of rational beings endowed with freedom, was clearly expressed by Origen in *On First Principles* (e.g. Bk II. IX : 6). But this doctrine of Origen has its counter-part in Origen's theory of the pre-existence of souls (IX : 7) which Gregory rejects. Origen discusses free will at length in Bk III, ch. I of the same work. His view is that pre-destination is the common sense under-

his philosophical dialogue *Contra Fatum*,[22] set in the form of the report of a disputation Gregory had with a heathen philosopher at Constantinople. He vigorously attacks the idea of pre-destination which sought to put the blame for the unbelief in the world on the decree of God and not on the will of Man. For Gregory, if that is what pre-destination means, then it is the chief sign of unbelief itself, to put the blame on God. The idea of pre-destination belongs to Hellenism, not to Christianity, Gregory argues very forcibly.[23] It is only the unbeliever who can say, if God wants me to be a Christian, I will become so even if I do not will it. Gregory's basic rebuttal is that Fate thus becomes a substitute for God, an arbitrary ruler of the world who allows no freedom in it. The pagan philosopher of Constantinople reminds us very much of our contemporary Jacques Monod's *Le hasard et la nécessité*[21]

Gregory's argument against astrology and horoscopes need not detain us. If the congruence of particular stars and planets at the time of a child's birth absolutely determines his destiny and character then two infants born at exactly the same time should have exactly the same destiny and character. At the time when a tyrant or an emperor is born, there are other infants born. But these latter, we know, do not become tyrants and emperors. Many people die together in an earthquake or war. They were not all born the same instant. Different nations have different customs. How can the influence of the stars have caused these differences, if human choice had nothing to do with them? The occasionally correct instances of astrological prediction are too rare to prove them to be anything more than the result of accident or chance. One could give instances or at least as many predictions that have gone wrong. If occasionally they appear miraculous, it is by demonic power that such knowledge is attained.

---

standing of the Scriptures, and has to be rejected if it is to mean that our wills are totally powerless to choose between good and evil as Augustine was to have said later. For detailed study of Origen's view on Freedom and Free will, see, G. Vinard, *Etude historique de la liberté humaine chez Origène*, Angers 1911; E. Maurial, *Origenis de libertate arbitrii doctrina*, Montpellier 1956; for full bibliography see, Henri Crouzel, *Origène et la "Connaissance Mystique"*, Desclée de Brouwer, 1961, p. 568.

22. *Kata Heimarmenē*. PG 45. 145 C-173 D.

23. *op. cit.* PG 45. 148 A-B.

24. Seuil, Paris 1970.

Dom Amand points out that most of these arguments are not original with Gregory, but taken from Karneades of Cyrene (214—129 B.C.), the Sceptic, founder of the Third (New) Academy, powerful opponent of Stoicism.[25]

The theological argument of Gregory, which we have already cited from the *Oratio Catechetica*, is that for God to force man either to belief or to unbelief, would be a violation of the freedom of man, which God Himself would not do, for his original gift of freedom is something about which he does not repent. For Gregory, an Augustinian or Calvinist doctrine of pre-destination (as distinct from that of St. Paul which occurs in a different context) and sovereign grace can belong only to paganism. It has no place in Christian teaching which regards Man as free because created in the image of God, and for which virtue that is compelled is not virtue.

The fatalism of Stoicism and Gnosticism comes from the concept of causality and unity of the cosmos (*sympatheia, synploké*), though as we have pointed out earlier, man can by his free choice contribute a final operative cause to the way things happen. But for the Stoics all this is implied also by the total immanence of God in the cosmos. God Himself is simply the logos that pervades the universe, and nothing transcendent. Man can conform to this logos, and accept the inevitable with calm and poise, by going along with the logos.

For Gregory, the absolute transcendence of God expressed in the concept of *diastēma* is the basis of God's freedom. God is not "affected" (that is what *pathé* means) by what happens in the cosmos, for while the cosmos is in time and space, God's Being is beyond the *diastēma.* God's transcendence thus is not simply a spatial removal beyond all that is in the cosmos, nor is it, as some contemporary theologians following Ernst Bloch would have it, the transcendence of the future. It is freedom from necessity, the freedom from being forced, compelled, affected, drawn, constrained by something outside of oneself that constitutes God's transcendence. It is in this sense that God's *apatheia* is to be understood, not

25. David Amand, *op. cit.*, p. 431. Dom Amand proposes, "avec modestie", the hypothesis that Gregory used a literary source emanating from Karneades (Carneades) and his school. He detects the influence of Karneades also in the *Oratio Catechetica* ch. 31.

merely as the absence of animal passions. Transcendence is the basis of sovereignty. The *huperkeimenēs* is the *eleutheros*, the *autokraton,* the *autexousion.*

To be self-governing, self-regulating, self-authoritative, self-propelling, to be not the effect of outside causes, but a new cause which creates effects outside itself for the good—that is the essence of both freedom and of transcendence—not spatial separation or temporal potentiality. It is this quality that God has given to man as his most precious gift, the greatest value that makes man "lord of the earth."

Once we see the centrality of this notion of freedom, we can see why Gregory's thought cannot be apprehended in the categories of Nature and Grace, or Nature and Supernature, or Justification by Faith as understood by the Reformers, or Pre-destination and sovereign Grace.

3. God's Freedom in Immanence

We have dealt with God's immanence sufficiently in the previous chapter, as to show that the creation has come to be, is subsisting, and is moving towards its destiny, depending only on the will of the Creator. Whereas the Stoic imprisoned God in the cosmos as the soul in the body, Gregory removes the *ousia* of God completely from the creation. But the creation is not outside of God—in so far as the *energeia* which created and sustains it is not external to the *ousia* of God. *How* it is in God, cannot be conceived by man, as Gregory never tires of saying. He denies that the cosmos is related to God in the same way as the soul is to the body.

> "The question of 'how' has to remain beyond the reach of our curiosity. For even the fact that the visible came out of that which is in no way manifested, we have learned only from faith, as something which is beyond our capacity to understand."[26]

But that nothing can exist without God's will, Gregory never doubts. To quote just one passage at random from *On Infants'*

26. *De hominis opificio.* XXIII. PG 44. 209 C. See also PG 46. 121 B. NPNF V. 413 B. See also V. 457 B.

Early Deaths, about looking for a justification for the fact that some children die early :

> "I would say that it will be better not to dispute this matter in a rhetorical manner, fighting with a chain of extemporaneous answers to objections raised, but imposing some kind of order for the discourse, we should seek a proper understanding of this problem in a sequential and logical manner. What is that order? First, to know where human nature comes from, and for what purpose it came to be. If in these questions our answer does not go wrong, we shall not be far wrong in the answer to our question also. It would be superfluous to demonstrate logically that all that exists after God, whether intelligible natures or sensible, so long as they are, they have their being from God. No one, I believe, with some knowledge of the reality of existents would dispute this basic assumption. For it is acknowledged by all that the All is dependent on a Single Cause, and that none of the existents has its existence from or in itself, nor has it the source and origin of its existence in itself. There is, on the other hand, one Uncreated and Eternal Nature, ever the same as itself and existing in the same manner, transcending all dimensional conception, never increasing or decreasing, understood to be beyond all definition and limit, whose work is Time, and Space, and all that are in these, as well as anything before Time and Space, such as the mind conceive the intelligible and super-terrestrial world to be. One of these things which came to be, we say, is human nature, this being confirmed by some word of the inspired teaching, which tells us that after all things had been brought to birth by God, Man also was manifested upon the earth, with a nature commixed from heterogeneous elements, the divine and intelligible nature commingled with that portion of human nature contributed to it by the various physical elements, that thus Man came to be alongside his Maker, in order to be an intelligent animate statue of the Divine and Transcendent Power."[27]

The creation, as well as Man who is in it, are thus from God, in

27. *De Infantibus.* PG 46. 172 B-D. NPNF V. 375.

God, contingent upon God's will as the ground of its being. Gregory's thought should not be confused with the Tillichian idea that God is simply "the ground of our being". It is not God's *ousia* that is the ground of our being, but His *energeia*, His will and wisdom as it operates in creation. As Gregory says in his reputation of Eunomius, apropos of the scripture text which says: "God said, let there be light":

> "The words 'he said' do not refer to the voice and spoken word of God, but express to our sense-bound minds, the meaning of the idea that God's power is concurrent with His will. For all things consist in the will of God (*thelēmati tou Theou sunestē ta panta*), it is the habit of men first to express the will by word and thereby bring the will into effect, the scriptural account of the creation of the world is a sort of beginners' introduction to the knowledge of God, presenting the power of God's nature through matters easier to understand, for sense perception is a kind of preparation for the knowledge of intelligible reality, it is for this reason that Moses, through saying 'God said', presents to us the motive power of God's will as the precondition of all coming into being, and by adding immediately, 'and it was so' shows us that in the Divine nature there is no difference at all between will and effective operation (*mēdemian einai diaphoran boulēseōs kai energeias epi tēs theias phuseōs*), teaches us that the purposing (*hē noēsis*), is the principle of the operation (*tēs energeias*), the effective operation in no way falling short of the purpose, but the two are to be understood together at once, i.e. the movement of the mind and the effecting power of the thing purposed. For there is no intermediate step to be conceived between the word of purpose and the effect, as the shedding of light coincides with the kindling of the flame, both coming from the flame and shining forth along with the flame, in the same manner, the hypostasis or subsistence of all that has come to be is the effected work (*ergon*) of God's will, not something which happens in a second order after the will."[28]

This then is God's freedom, and the basis of both the subsistence

28. *Contra Eunomium* II. 227-229. GNO I. 294/4-28. PG 45. 985 D-988 B. NPNF V. 273 A-B.

of the creation, and the immanence of God in creation. His will becomes the creation. The creation is present to God as His will is present to Him—not as something external to God, but "in" God, as the will is "in" a person. The idea is even more clearly expressed in *The Hexaemeron,* Gregory's commentary on the account of creation in the first chapters of Genesis. Let us repeat the whole passage we have cited before in brief:

> "It is necessary, it seems to me, before setting forth the contents of the Scriptures, to agree on the following principle: In God's nature the power is concurrent with the will, and the measure of God's power is the will (*epi tēs theias phuseōs sundromos esti tēi boulēsei hē dunamis kai metron tēs dunameōs tou theou to thelēma ginetai*). But the will is wisdom. And the property of wisdom is not to be ignorant of how every single thing has come to be, for the power is connatural with the knowledge (*tēi de gnōsei sumpephuke kai hē dunamis*). Simultaneously with his knowing that something ought to exist, immediately the power to bring existents into effect co-functions with the knowledge, conducting the known thing into effective operation, and the effect is in no way different or lacking from that which is known, rather, together at once, without any interval, the work or effect is comanifested with the will. For the will itself is thus power, such that whenever existents are willed to come into being, the opportunity is also provided for the willed beings' existence. Thus all at once, synchronously, have all things to be conceived in God in relation to creation—the will, the wisdom, the power, and the being of the existents."[29]

Thus God's will becomes concretized as the creation.[30] It is His freedom, wisdom, and power that gives birth to all existents.

---

29. *In Hexaemeron.* PG 44. 68 D-69 A.

30. To cite only a few further examples of this idea in Gregory see the work on "*When all shall be subjected to him*". PG 44. 1312 A. *to theion thelēma hulē kai ousia tōn demiourgēmatōn egeneto.* Or the clear statement in the *Life of Gregory Thaumaturgus* (PG 46. 920 A). "If so great is the power of the servant (as to perform all miracles) . . . how much superior we must conceive the power of the Lord of All to be, whose will alone has become both the matter and the shaping, as well as the dynamic power of the cosmos and of

They also continue in existence only by His will sustaining them. Thus God is not caught in His cosmos, but remains master of it, knowing it, willing it, empowering it, leading it to its fulfillment. The creation is not free from God in the sense that it can exist without him. It is only free in the sense that certain free beings in it have been given the freedom to accept or refuse the existence given to them. They can choose to refuse; this would be a choice of non-being, choice of evil, for the existence given to them is a good existence, but not compulsorily good.

But God's *ousia* is not dependent on the creation, and nothing in creation has a "part" in the *ousia* or *phusis* of God—not even the human soul.[31] God's immanence in the creation is not by *ousia,* but by *energeia.* His freedom then consists in the capacity to bring into being, sustain and lead to its fulfillment, the creation. He is not compelled by any force to do it, but this is His goodness as it expresses itself in freedom.

Jerome Gaith has taken note of this as an important aspect of "La Liberté Divine",[32] in a manner that goes much deeper into Gregory's thought than Dom David Amand was able to. The present author acknowledges his deep indebtedness to Gaith's seminal work which has received only scant attention in the western theological world with its centuries-old fear of freedom.

But the question remains, how Man who is the image of such a powerful Creator, whose will and word not only become law, but also *fact* in the very act of conceiving or willing, can participate in this kind of power. It seems absolutely necessary that man should have some part in this kind of power, if he is truly to be a visible image of the invisible God. For power is indeed a central

---

all that is in it and above it! (*hou to thelēma, hoion tis hulē, kai kataskeuē kai dunamis autou te tou kosmou kai tōn en autoe pantōn, kai huper touton ontōn egeneto*)".

Or again in the *Oratio Catechetica* 24. Srawley 91/10-14. PG 45. 64 C. NPNF V. "And let it not sound strange to some ears if we say that the whole creation in the cosmos and all that is understood to be outside the phenomenal world, have their subsistence in God, His will in some way having been transformed into the being of those that pleased him. (*autou tou thelēmatos pros to dokoun ousiōthentos*)".

31. *De Anima et Resurrectione*. PG 46. 121 D. NPNF V. 458 A.

32. *La Conception de la Liberté chez Grégoire de Nysse*. Paris 1953, pp. 31ff.

aspect of the glory or the goodness of God, and if Man does not participate in that he can hardly be called a faithful image of the original.

We have already referred[33] to the fact that Gregory of Nyssa regards the miracles wrought by Gregory the Wonder-worker as manifestations of the latter's participation in the creative power of God. Man's freedom to affect the world by his will and to cause certain effects (always for the good) in the natural order by willing it in prayer, was what the Thaumaturgus expressed in his life.

Just as the Transcendence of God means that He is not simply a first link in the chain of Being, but the source and origin of all being without being affected by that which He has created, so also Man has both to remain in *apatheia,* unaffected by causes outside of himself, which is his transcendence, and at the same time he has to be *immanent* in the creation by his presence in it, transforming it by his will, by his prayer, by his *energeia* or operations.

Man's presence in the material creation is what lifts it up to God. It may be as a result of sin that we are in this heavy body in this material world dominated by the downward-pulling force of gravity. But man's presence in matter is of the utmost consequence to matter itself. Gaith calls it "immanence consciente"[34]—a conscious immanence, an immanence of God through a conscious, rational, free being who is part of the cosmos, who is also ubiquitous in it, capable of shaping it, moulding it, transforming it, humanising it, and lifting it up to God.

God is present in a mysterious way in the soul of Man united to Christ and separated from evil, so that Man can see in himself, when purified, incorporated into Christ, and cleansed from all evil by the Spirit and by the life of virtue, the reflection of the original goodness.[35] The Incarnation of God in Man effected in Jesus Christ is thus for Gregory the fulfillment of the vocation of Man to be God's presence consciously immanent, in creation.

Balancing the Transcendence of God through the *diastéma* there is thus the Immanence of God in the creation through *metousia*

---

33. See note 30 above.

34. *op. cit.*, p. 36.

35. *De Beatitudinibus.* Oratio VI. PG 44. 1272 C. *From Glory to Glory*, p. 102.

or participation. That *metousia*, a possibility given by the act of creation, becomes realized in the act of Incarnation as a historical reality.

4. Freedom as Creativity of the Good

God's freedom is thus both in his *ousia* which lies beyond the *diastéma*, and also in the *energeia* or operation, from which the creation takes its source and by which the creation is led forward through time and space towards the good. The free *energeia* which proceeds from the free *ousia* is the manifestation of God's freedom.

So also for Man. His *ousia*, by being united to Christ and by separating itself from evil, attains *apatheia* or transcendence from external necessity and compulsion. But this transcendent and free *ousia* or *phusis* of man has to manifest itself through man's operations. This operation of man in freedom is what is called *virtue* by Gregory.

The word virtue is misunderstood in our modern languages as mere absence of evil. This is only the negative aspect of virtue. Virtue is a power, a *dynamis*, that results in actions. This was the way Plato himself had conceived virtue.

Plato had given the definition of virtue as proper active functioning—"the function of a thing being that which only it can do or that which it does best".[36] But we should not overlook the fact that pagans often spoke of the *aretē theou*, the virtue of God.[37] It is in this context that we can understand I Peter 2:9 where part of the task of the Christian community is to "declare the virtues (*tas aretas*)"[38] of God who called them out of the darkness of evil, to the marvellous light of walking in God's ways. God's virtues are his mighty acts—his operations or *energeia*, by which he rescued Israel from Egypt, or raised Christ from the dead.

Man if he is in the image of God and free, has also to practice the same virtues—i.e. mighty, heroic acts for the good of others, not because an external necessity compels them, but because it is their nature to do mighty acts in freedom.

---

36. *Republic*. Part I. Bk I. C. See W. Theiler, Zur Geschichte der theologischen Naturbetrachtung bis auf Aristoeles, p. 74ff.

37. e.g. in several pre-Christian inscriptions. See Liddell & Scott s.v. *aretē*.

38. In the Latin Vulgate—*ut virtutes adnuntiatis ejus*.

So the way in which Man uses his *energeia* is determinative for the expression of his *ousia*, as it is in the case of the prototype. Man can choose to use it to be an animal; then he chooses that kind of being. Or he can use it for *virtue*, mighty, heroic, good acts, then he more and more appropriates his true being.

To seek pleasure instead of joy is to choose the animal instead of the divine, as determinative of the being of Man. *Philhedonia*, love of pleasure, is the opposite, the counterfeit substitute, not of pain, but of Beatitude (*makariotēs*). Temptation means to be placed on the middle ground between pleasure and joy, so that man can freely discern and choose what is good, and by that choice determine his being to that extent. Some people learn the hard way, the way of experience, by tasting pleasure and only later finding out its emptiness, says Gregory.[39] But not all can afford to do so, "for it is not by our own experience, that we learn everything"[40] we can already know these pleasures from our own reasoning (*dia monōn tōn logismōn*) since our poverty (unlike Solomon) prevents us from enjoing these pleasures to the limit. Solomon reached the limit of enjoyment and discovered the pleasures to be empty. Solomon did not even have to fight his way to kingship. But since there was no limit to the pleasures he could enjoy, he tasted all the pleasures, and because he was also wise, discovered and declared that they were all empty. But once you fall into the vicious grip of these pleasures, you may still be able to bare your teeth and laugh, but inwardly you are gloomy and unhappy. The reason is no longer free, the will simply follows wherever the drives of the body drag it, and man has become an animal. He is no longer free to "conduct his operations to a freely chosen goal" (*mēte ergon epi tini skopoē katorthoumenon*),[41] and has thus lost his humanity and become *alogon*, an animal.

Another fine flourish of Gregorian rhetoric occurs in this context of describing the folly of hedonism, in the commentary on *Ecclesiastes*. The pursuit of passion or pleasure as a possession is folly, madness, paranoia (*paraphora kai paranoia*).[42] Passion and the pursuit of pleasure as a possession can only be a denial of

39. *In Ecclesiasten* II. GNO V. 306-307. PG 44. 644 A-B.
40. *op. cit.* GNO V. 306/3-4.
41. *Ibid.*, GNO V. 310/10.
42. *Ibid.* GNO V. 310/8-9.

humanity and a descending to the animal stage, a selling into slavery of one's own soul.

> "Why do you soften the intensity of the mind thus? Why cut the sinews of the tonality of the soul? Why maltreat the reasoning capacity? Why. tell me, do you darken the purity and clarity of your mind by this gloom that is outside of its nature?"[43]

If passion thus is to allow the drives of the body to rule the will and enslave it, virtue is its opposite—namely directing the body to the good by free choice of the will. The will is creator of one's own nature by free choice. This possibility is part of the freedom of man, a participation of the *autexousia* that belongs properly to God and is His gift to Man created in His image. We can refuse that image by choosing to conform ourselves to the image of the animal. For passion is what makes us akin to the animals. In the animals the passions serve the purpose of self-preservation and continuation of the species. But transferred to man, they became the counter-force against which he has to establish his freedom and shape his own nature.

Gregory in *On the Making of Man,* compares Man to those extraordinary pieces of Greek art in which two faces are combined in one form:

> "Thus Man, it seems to me, exists in a double likeness to two opposite things, on the one hand, through the Godlike mind and will conformed to the Divine Good, on the other hand drawn to the likeness of brutes by the passionate impulses that arise in him. In many ways his very reason is rendered brutish by the inclination and disposition towards the irrational, the better element obscured by the worse. For whenever anyone by his own free choice drags down the mental operation (*ten dianoētikēn energeian*) to these irrational passions and compels his reason to be a subordinate of the passions, some kind of transformation of the imprint of the good towards the image of the irrational takes place (*paratropē tis ginetai tou agathou charaktēros pros tēn alogon eikona*), the nature itself being tra-

43. *Ibid.* GNO V. 311/11-14.

ced in this pattern, as if reason itself were the cultivator sowing the seed of the passions and little by little tending them to full growth, for once having lent its co-operation (*sunergia*) to the passions, he raises an abundant and thick-grown crop of these absurdities. Thus the love of pleasure (*philēdonia*) becomes the beginning of the likeness to the irrational brutes, and its growth comes from the many human transgressions, and gives birth to the several sins of pleasure which are not to be found among irrational animals. Thus for example the rise of anger (in man) is of the same nature as the impulse in the irrational brutes, but when reason joins the fight on the side of the passion, it becomes stronger. Thence arise malice, jealousy, falsehood, conspiracy, hypocrisy. All these are cultivations of the evil mind. If the passions were deprived of the alliance in battle of the reason, anger remains something short-lived and unintense, bubbling up into existence for a moment, to be blown up into nothing the next. Human covetousness is the greed of the swine introduced into his nature, human arrogance the high spirit of the horse. All these characteristics of the brute creation provide the potentiality to become evil through their evil use by the mind, just as, on the other hand, if the mind can resist the power of these impulses, each of these very impulses can be transformed into virtues. Then anger can make courage, cowardice can produce caution, fear can become loyalty, hatred be directed to turning away from evil, the impulse to love can be made the yearning for the Truly Good. The exaltation of character rises above the passions and the will is conserved from submission to evil."[44]

Thus, for Gregory, though human nature is created by God, it is only the potentiality and the freedom to create oneself in accordance with one of two different patterns—animal or divine—that

44. *De hominis opificio*. XVII. PG 44. 192 D-193 C. NPNF V. 408.

The idea of this double tendency, which is the milieu in which human freedom has to be exercised in history, occurs practically in all of Gregory's works. To cite only a few examples:

De Virginitate 4. GNO VIII. 276-277.

In Psalmorum Inscr. II. 6. GNO V. 88/5-9.

In Cant. Canticorum 12. GNO VI. 345/9—364/2.

God gives. The capacity to create oneself, so to speak, is part of the image. Even as God has no other source of being, it is given to man, as a gift from the Creator, the freedom to create himself, to be a coworker with God in perfecting His creation.

Christian theology has been generally reluctant to accept this idea of Gregory's—that Man is not simply a creature pure and simple, but a co-creator of himself and his world. Only Teilhard de Chardin, among recent thinkers, has developed this idea as a central element of his thought. But our own experience during the centuries has shown us in what large measure "Man makes himself," for the good or for evil.

To see the human enterprise as a joint operation between God and Man is neither dishonouring God's sovereignty nor exalting Man above his created limit. It is in this context that both the notion of virtue as rectification and the free cooperation (*synergeia*) of man as a necessary element in God's plan for the creation have to be understood in Gregory's thinking.

Gregory defines virtue as the rectification of the faculty of freedom:

> "Who does not know that Virtue is the rectification of the faculty of choice? And the flesh is the instrument of choice, impelled by the effort of the understanding, being transformed into that towards which the motive force leads it. The faculty of choice is thus nothing else than the mind and the disposition towards something."[45]

For Gregory the mind (*nous*) includes understanding, imagination, projection, purpose and will. All these are in need of rectification, and virtue is the process of straightening out the mind-will through discipline and the practice of the good. It is in this process that Man becomes fully man, partly self-made.

But then what has happened to the doctrines of Sin, Grace, and Incarnation? Is this not a doctrine of salvation by works? To this question we turn in the next chapter.

---

45. Antirheticus adversus Apolinarium. GNO III. I. 198/1-5. PG 45. 1217 C. (*Tis gar ouk oiden hoti aretē prosaireseōs esti katorthōma, hē de sarx organon esti tes proaireseōs, tēi hormēi tēs dianoias egkeimenon, pros hoper an ho kinōn agēi, pros touto metatithemenon. Hē de proairesis ouden heteron ei mē nous tis esti kai prós ti diathesisl.*

CHAPTER VII

MAN AGAINST GOD
Man in History

Gregory was not a naive humanist who simply exalted human nature in his lofty and noble conceptions. He could, when necessary, come out with a very rhetorical downgrading of Man. In his argument with the Macedonians (who denied that the Holy Spirit was God), he accuses them of bringing down the glory of God by seeking to sit in judgment on God's nature, weighing one divine person over against another, judging the Holy Spirit to be of less value than the Father and so on. For however high the human mind rises, it cannot ever catch sight of the "footstool"[1] of God's *ousia,* without being struck with awe and wonder. Compared with God, Man is nothing:

> "If then even the highest of human powers are below the majesty of Him who is worshipped (for that is the metaphorical meaning of the phrase the footstool), what vanity is theirs who think they have within themselves the power to define and set limits to the value of the honour and worthiness of the nature which is beyond all evaluation, and thus to judge the Holy Spirit whose power is far superior to their powers to be unworthy of some of the attributes of honour, and to measure and separate the worthiness of the Spirit ! O ! what wretched and miserable madness ! How little they understand who they are who presume to say such things and who the Holy Spirit is, against whom they rage so arrogantly! Who will tell these people that men are but 'a wind that bloweth and returneth not', built up in their mother's womb, sordidly put together, and sordidly to be dissolved again into the earth, whose lot is

1. Ps 98(99): 5.

> a life like that of grass, blooming for a while through the illusion of life and then withering away, all the blossoms also falling around them and disappearing, not knowing exactly what they were before birth or into what they will be changed, ignorant of the very destiny of their own souls while they are in the flesh! These human beings!"[2]

1. Sin, Original Sin and Nature

Gregory takes human sin seriously. Sin is separation from God, the source of all being, and therefore, death.

But Gregory denies some affirmations of our later theology—(a) that human *nature* is sinful; (b) that *all* men are equally sinful and (c) that man can do *nothing but* sin.

Human nature cannot be sinful, for nature is what is created by God, and it was not created evil or sinful. What is constitutive of our nature is that it was created in the Image of God, who is the perfection of all goodness. But precisely because freedom is part of the image, the created nature has to be "worked out" through human freedom. There are two possibilities open to man—one, to say "yes" to the existence given to him by affirming that it comes from God and by working out in freedom its true nature as good. This is life. The other possibility is to say "no" to that existence by refusing to acknowledge that it comes from God, thinking it is one's own and by refusing to work it out as a manifestation of God's own glory. This is *alloiosis*, alienation, death.

What is the origin of sin in humanity? Gregory does not question the scriptural account in its essentials. Sin came by "eating", by an act of being drawn to the "Tree of the knowledge of good and evil" through external persuasion and actual tasting of the fruit of that tree.[3]

But what is this tree, and what is this "eating"? In a piece of rather sophisticated exegesis, Gregory argues that the sin of Eve was not simply in choosing evil with the knowledge that it was evil, but in choosing something which seemed good, but was not good for her. The tree of the knowledge of good and evil could not have been evil, for it was created by God, and everything God created was necessarily good.

---

2. *De Spiritu Sancto.* GNO III. I. 107/29-108/17. PG 45. 1328. NPNF V. 323 B.

3. *De hominis opificio.* xx. PG 44. 197 C-201 A. NPNF V. 410.

But it was not good for Eve, for it had been forbidden. She could have, if she had adhered to the word of God, seen without actually tasting it, that it was not good for her. But the enemy persuaded her that it was good for her, and she adhered to the word of falsehood. If it had appeared as pure evil, she would not have tasted it. Only because it had the appearance of Good, capable of making men "like Gods", which she must have rightly judged to be a good thing, she had a wrong discernment of what was good for her at that time. So while the tree and the fruit are not evil in themselves but good, they became evil to her, in so far as falsehood was the basis on which she approached it.[4]

The "knowledge of good and evil", as Gregory exegetes it, is thus being drawn towards something which one wrongly or falsely judges to be good for oneself, on the basis of false belief and desire for gratification rather than for the sake of the good itself. Evil is a latent possibility in the misuse of the good. It was this *misuse* of the good that constituted the first sin. So the fruit of the tree of the knowledge of good and evil, good in itself, and good for man and woman also if one refrains from eating its fruit, thus can be misused by eating it against the commandment. Gregory says :

> "Since he foresaw this possibility, the Serpent points to this evil fruit of sin, not as having evil as its nature, or manifestly appearing as evil (for then man would not have been deluded into choosing manifest evil), but decking the phenomenon she saw with a glamour and conjuring up in her taste the potential pleasure of sense-experience, he (the serpent) appeared convincing to the woman, for as the Scripture says: 'And the woman saw that the Tree was good for food, pleasing to the eyes, and ripe for knowledge, and she took the fruit thereof and ate.' But this food has become the mother of death to human beings."[5]

Sin is an act of the will, not the bodily drives or the surge of the passions. Gregory leaves us in no doubt what the source of evil is—the freedom of Man. Eunomius seems to have implied that the body is the source of sin, while the soul remains sinless. He seems

4. *De hominis opificio*. xix. PG 44, 197 B. NPNF V. 409 B.
5. *De hominis opificio*, xx. PG 44. 200 C—D. NPNF V. 410 B.

to have expressed the view that only the body of man had to be saved, and therefore that the Logos assumed only a body, but not a human soul. Gregory quotes from Eunomius the sentence:

> "He who in the last days became Man did not assume manhood composed of soul and body."[6]

The denial of a human soul in the Incarnate Christ was intended to show that the subordinationist words of Jesus like "My Father is greater than I" were not spoken by the human soul, but by the Logos himself. Gregory calls this a monstrous distortion of the teaching of the Scriptures, and enunciates the central principal of Church teaching that

> "He who was tempted in all points as we are, yet without sin, left nothing of our nature unassumed."[7]

For Gregory thus it is clear that if the Incarnate Logos assumed the whole of human nature, not leaving anything of it unassumed, then sin is not an integral part of that nature. Gregory would thus insist categorically that "all sin is against nature" and therefore that sin cannot be a part of nature.

The body is not sin; neither is the soul sin. (*hē de psuchē hamartia ouk estin*), but being ill-advised became capable of sin (*alla deklikē hamartias et aboulias egeneto*).[8] Sin is an act of the will, and the will is an activity of the soul:

> "For the disobedience of the faculty of choice is not a sin of the body, but the faculty of choice is a property of the soul, from which the whole misfortune of our nature takes its origin, as witness the words of the unlying warning of God, that on the very day when they would lay hold of that which

---

6. *Ton ep' eschatōn tōn hēmerōn genomenon anthrōpon ouk analabonta ton ek psuchēs kai sōmatos anthrōpon.*

Refutatic Conf. Eunomii. 72. GNO II. 384/29-385/2. PG 45. 545 A. NPNF V. 126 B.

7. *kai ouden aphēke tēs phuseōs hēmōn, ho ouk anelaben ho kata panta pepeiramenos kath' homoiotēta chōris hamartias.*

Ibid. 175. GNO II. 386/9-11. PG 45. 545 C—D. NPNF V. 127 A.

8. *Ibid.* 176. GNO II. 386/11-13.

> is prohibited, forthwith along with the eating, death would also be laid hold of."[9]

Gregory goes on to argue that since Adam continued to live on in the body for many hundred years after his sin, the death that ensued immediately on the eating was not the death of the body, but a sentence of death against both body and soul.

What then does the death of the soul mean? It does not, for Gregory mean the simple coming to an end, for if the souls had simply come to an end at the time when Adam and Eve ate the fruit, their bodies could not have lived on. Death, for Gregory, is essentially separation from the source of life, which is what sin also means:

> "For sin is estrangement from God, who is the True and the Only Life."[10]

Estrangement or alienation from the life of God immediately entails the sentence of death from the moment of separation, though the life of both body and soul may in fact continue for some time.

Gregory does not believe in any kind of eternity of the soul. It begins with the body, and dies with the body. Here he makes a bold departure from Origen, who somewhat along Neoplatonist lines, conceived the soul as pre-existing the body.[11] Methodius in

9. *Ibid.* 174. GNO II. 385/13-18.

10. *Ibid.* 174. GNO II. 385/24-25. *Hamartia de estin hē tou Theou allotriōsis, hos estin hē alēthinē tē kai monē zoē.*

11. Our sources for precisely fixing the teaching of Origen on the pre-existence of souls are extremely meagre and quite unreliable, having been taken either from Jerome's and Ruffinus' indirect reporting in Latin, or from the anathemas of later councils which condemned Origenism by formulating it into easily condemnable formulas. That Origen did advance the theory of the pre-existence of all souls, however, is extremely credible. See Koetschau's Text of *Peri Archon* Bk. II. chp. VIII and IX. On the problems of this text see Franz Heinrich Kettler, *Der ursprungliche Sinn der Dogmatik des Origenes,* Berlin 1966, especially the section II on the difference between dogma and hypothesis, pp. 12ff. Origen says, according to Jerome, that "haec juxta nostram sententiam non sint dogmata, sed quaesita tantum", and according to Ruffinus, that "sed haec, quantum ad nos pertinet, non sint dogmata, sed discussion is gratia dicta sunt."

Kettler *op. cit.*, p. 16.

reaction against Origen went to the opposite extreme, basing himself on a literal interpretation of the text in Genesis which said that God first fashioned the body and only then breathed the soul into it; the soul thus being created after the body, as a kind of seed placed inside the dead earth of the body.

Gregory cites both Origen and Methodius in an inmistakable allusion to *On First Principles* (*Peri Archōn*), he attacks the view of Origen (without specifically mentioning his name, but only alluding to the title of Origen's work) about the pre-existence of souls, as a still on-going debate in the churches of his time. He refers to "those who before our time worked out a discourse on First Principles"[12] and spoke of a pre-existing community of souls. He refers also to Methodius;

> "Others, however, taking the order of creation according to Moses, say that the soul is second in rank to the body."[13]

In a vigorous refutation of Origen, Gregory accuses him of "absurd reasoning", an "incoherent fable", lacking beginning or conclusion. He rejects also the doctrine of Methodius, and wants to ply a middle course, by asserting that the soul and the body were created together, that they have one beginning.[14] Both have to grow from small beginnings to fullness of form by change. The soul is no more changeless than the body. Here again Gregory radically departs from Platonic or Neoplatonic views about the unchangeable character of the soul.

It is in this context that Gregory sees man as born to a death-bound life. It is from Adam and Eve, whose bodies and souls were condemned to die, that our bodies and souls are derived. They are thus born mortal, subject to death.

But here one has to make a very delicate clarification. It is not sin that we inherit from Adam by nature, but the consequence of sin, namely mortality and corruptibility. At this point Gregory and the Eastern tradition in general, would find the Augustinian notion of "original sin" unacceptable.

In fact there exists no such terms as "original sin" in the earlier Eastern patristic writings, either in the Greek or in Syriac. The

12. *De hominis opificio*. xxviii. PG 44. 229 B. NPNF V. 419 A.
13. *Ibid*.
14. *Ibid*. xxix. PG 44. 233 D—236 B.

idea of "natural sin", advanced by some, was vigorously refuted by the great teacher of the Syrians, Severus of Antioch, who has been much misunderstood and more mis-interpreted in the West. Severus who wrote in Greek but is preserved for us mainly in Syriac translations, found the idea of "natural sin" (*Htito Kyonoyto*), which must have been *hē phusikē hamartia* in Greek, absolutely unacceptable to an orthodox Christian. Severus was preaching in the cathedral of Antioch against a contemporary heretic (Romanos of Cilicia) who had suggested on the authority of St. Paul that real sin was "bodily sin" which is also "natural sin". The heretic seems to have made the distinction between "natural sin" and "voluntary sin". Severus argues quoting Chrysostom, Nazianzen and Cyril of Alexandria (but not Nyssa), that if bodily sin is natural sin, and is the only real sin, then voluntary sin must be unreal. Severus calls the distinction unscholarly. Sin is sin because it is voluntary. Otherwise God would not condemn us for it. "It is a disease of the will, and disease is not natural."[15]

All sin is against nature. This is a fundamental principle in Eastern patristic thought, and Gregory has philosophically made clear why this is so. Human nature has become enslaved to sin and through sin to evil, but sin is an alien master that now rules man, not something that belongs to his nature. Gregory makes this clear in his commentary *On the Inscriptions of the Psalms.* The source of sin is in Man's changeability which is also the arena of his freedom and self-creation. In fact Moses, the Man of God, in Gregory's very human exegesis of the title of Psalm 89 (which reads in the LXX, A Prayer of Moses, the Man of God)[16] pleads with God that while He (God) Himself is unchanging and unchangeable, we (men) who are created changeable are in a very difficult situation.

> "For whereas the human has become dominated by the evil of sin, and being enmeshed in the contrarities of passions became broken off from the union with the good, and was in need of

15. Severus of Antioch. *Cathedral Homily CXIX Patrologia Orientalis.* Tom. XXVI. Fasc. 3, P. 392 ff. For a full discussion of the origin and development of Syrian thinking on Nature and Grace, see Arthur Voobus, *Theological Reflextions on Human Nature in Ancient Syrian Tradition*, in P. J. Hefner, ed. *The Scope of Grace*, Philadelphia 1964, pp. 99-120.

16. *Ibid.*

> an intermediary powerful enough to recall it from destruction, the Man of God comes to be for this mediation, pleading for mercy on behalf of his own people's faults, seeking to make God abashed enough to show mercy on the perishing ones. So as if pleading their cause in a court hearing, he (Moses) says that it belongs only to God to remain steadfast in the good always, unmoved, ever the same, but it is not given to human beings, located in change and alteration, ever to remain the same, without either being drawn higher to the good, or falling away towards the lower from the participation in the good, because of which he pleads that the straying ones may be made worthy of salvation in all their generations, in the immutable refuge."[17]

Human nature, for Gregory, is not sinful in itself, but ever under pressure to change, either for the better or for the worse. The need is to reverse the direction of the change; not to become changeless, which is impossible for any created nature, but to be redeemed from slipping down the path of evil, to be set up again on the upward climb into the infinite good. He continues, a little later:

> "For He said: 'Return, o sons of men'. What is the teaching here? The word refers to the nature, and implies the healing from evil. For since being changeable you fell away from the good, you need again to be changing for the good. And wherefrom you fell away, towards that return. Thus it is in the choice of men to conduct themselves to that which they wish, either to the good or to the evil. What follows is another teaching. For it says 'For a thousand years are in Thine eyes, Lord, as yesterday, a day which has passed, their passing into nothingness (*exoudenōmata autōn*) becomes the years (Ps 89:4-5). What shall we learn from these teachings? That for him who returns again to the good from his turning away, even if his life is spotted with a myriad faults, the multitude of evils appearing together as a thousand years, when he turns to God it all becomes nothing. For before the eyes of God the present always remains present, never reckoned as having

17. *In Inscriptiones Psalmorum*. I. VII. GNO V. 45/20-46/6. PG 44. 457 C-D.

> passed away. As one day or as a part of the night it is judged by God even if it has passed or run along. To remain present in evil, even if disregarded as nothing by the sinner, can however be seen by God as the fullness of years, for it says 'their passing into nothingness becomes a year'. Well and admirably does it call the faults 'passing into nothingness' or 'annihilation' (*exoudenōmata*). For he who activates evil leads the sinner to nothingness, and to anyone who seeks to find excuses for each of his sins, it is legitimate to say 'Nothing is lust, and nothing is wrath, and nothing each such of sins'. For they are the movements of the nature. But the nature itself is the work of God. For Him who overseas the human existence, none of these (movements of the nature) are to be considered evil. So the Scripture says: 'These are passings into nothing'. But when the one who acts remains abiding in evil by choice and does not turn away from it, the evil acts remain before the eyes of God as a multitude of years."[18]

Gregory's teaching here is characteristically dialectic and fascinatingly sophisticated. Evil in our nature is "nothing", for it is not created by God, but is merely the movement of the created nature by its own free choice. It leads the doer of evil to nothingness. But when the man who has slipped away from the good and has become spotted with a thousand evils, turns to God in repentance, the nature of his past evils as nothing becomes evident in the presence of God, but instead of turning away from evil towards the good, if he persists in the evil, taking his excuse from the statement that evil is "nothing" because it is passing away, then that does not work, for that which is passing away for man, does not pass away for God, and all the evil is present before God and becomes the basis for the *exoudenosis* or annihilation of the sinner.

Evil is "nothing", only when the relation to the source of all good and of all being is restored through grace, repentance, and the separation from evil. Evil destroys human nature and reduces it to nothing when man persists in evil, remains separated from

18. *Ibid.* GNO V. 46/21-47/25. PG 44. 460 B-461A.

God and disregards sin as something unimportant and passing away.

Human nature is not evil, for it is God's creation. Human nature is still free to choose between good and evil, and that is the basis on which the call to repentance can be addressed to man. But the call is not merely in respect of each individual act, but for a continuing life separated from evil and re-united to God.

This is not a naive Hellenistic optimism about human nature. The wretchedness of human nature is a constant awareness in Gregory, and documentation is available in great abundance, especially in Gregory's epistles, for his low estimate of human performance. He does not, however, like Augustine, base his understanding of human nature on a phenomenological analysis of society or on an existential psychoanalysis of oneself. His view about the basic goodness of human nature comes primarily from the Judaeo-Christian revelation and not from Hellenistic anthropology, or personal psychological insights.

The same passage from the *In Inscriptiones Psalmorum* continues:

> "Again the transience of our nature is indicated by the Word, as if once again to put God to greater shame. For the next says, clearly bringing into view the picture of how the wretchedness of our nature should be regarded: 'morning and evening', i.e. youth and old age, 'the first green shoot springing up at dawn and then blossoming and then passing away'. And after this, our life-span, springing up from moisture, and blossoming on the plant, is dissipated into nothingness after the time allotted by nature, leaving a residue of dryness and wasting away. In this sense the word says: 'In the morning springing up as grass, in the evening blossoming and passing away, at dusk falling off, withered and dried up' (Ps 89:6). This is human nature. And in the next verse, he laments the state of humanity even further, saying that human life is consumed by wrath, as if torn asunder by some strong wind springing up from passion. For clearly by wrath and passion is indicated the self-sundering activity (*tēn apostatikēn energeian*), through which on the one hand human life is extinguished and on the other, its disorder comes to rest."[19]

19. *Ibid.* GNO V. 47/26-48/14. PG 44. 461 A-B.

The created nature has thus both possibilities, that of self-destruction and that of self-creating. This is its freedom. To adhere to evil is to choose destruction and non-being. To turn to God and to cleave to him, separated from evil, is to choose being and growth in the good.

The misery and grandeur of human existence is thus held in dialectical tension by Gregory, centred in human freedom and the grace of God. This is certainly not hellenistic, but can be traced back to the asian-semitic conception of human nature.

Professor Arthur Vööbus, the Syriac scholar, draws our attention to this basic semitic conception of Man in his illuminating article on *Human Nature in ancient Syrian Traditions.*[20] The impulse towards good (*yetser hatob*) and the evil impulse (*yetser hara'*) are both created by God, according to Rabbinic teaching.[21] This view is sustained in the early Syriac Patristic literature, with the important qualification that the impulse towards evil is a possibility of human nature seriously affected by the Fall, but what was created was only freedom, with the possibility of turning towards good or evil. We need to take seriously the contention of Vööbus that "for Aphrahat there is no original sin. Human nature is not affected by it."[22] Vööbus is right in saying that for the Syriac tradition, concupiscence is neither sin nor the consequence of sin, in itself. Sin has made it more powerful and impulsive, but concupiscence was only the occasion for sin.

Aphrahat cannot be considered normative for the Syrian tradition, in the same manner as Ephrem and Severus. For Ephrem, as Vööbus shows, the Fall of Man is a stark reality, resulting in Man's losing the "garment of glory", his body becoming coarser, pain, disease, toil, and suffering finding entrance into his life, with death as the final punishment.[23]

---

20. In P.J. Hefner, ed. *The Scope of Grace, Essays on Nature and Grace in honour of Joseph Sittler,* Fortress Press, Philadelphia 1964, pp. 99-120.

21. *Yoma* 69 b. *Midrash Bamidbar* R. 27: 8 .See article by Voobus on Aphrahat in *Jahrbuch fur das Reallexikon fur Antike und Christentum* III (1960), pp. 154ff.

22. *op. cit.*, p. 107. This is directly contrary to the view expressed on Aphraat (fl. 337-345) in the *Dictionnaire de Théologie Catholique* (Paris 1909, vol. I. col. 1463) by J. Parisot "Aphraat fournit de trés précieux témoignages en faveur du dogme du péché originel".

23. See Ephrem's *Carmina Nisibena* XXVII. ed. G. Bickell (Leipzig 1866), pp. 40ff.

We do not know if Gregory ever met Ephrem, or if they did, whether there was any interpreter present who could help their theological discussion, for Ephrem knew no Greek and Gregory knew no Syriac. The authenticity of the *De Vita S. Patris Ephraem Syri* attributed to Gregory in Migne's Text[24] remains to be proved. But whoever was the preacher of that homily, he claimed to have known Ephraim personally and regarded him as of the same stature as Basil, after both had departed this life.[25]

Vööbus cites Ephraim's saying, which well accords with that of Gregory: "If our created nature is ugly, the reproach falls on our Creator, but if our freedom is evil, the reproach accumulates itself on us."[26] Vööbus sees clearly that this affirmation of the freedom of man belongs essentially to the Semitic tradition. It is the strength and power of the human will that is basically denigrated in the Augustinian tradition. The Semitic tradition, on the contrary, affirms the will. The emphasis falls on how man responds to the grace of God and that is by will.[27]

We have already referred to Severus of Antioch's oppostion to

---

24. PG 46. 820-849. On the question of this meeting of Ephraim with the Cappadocians, see Dom Olivier Rousseau, *La rencontre de Saint Ephrem et de Saint Basile*, in *L'orient syrien* 3, Paris 1958, pp. 73-91.

25. PG 46. 821 D—824 A.

26. *Hymni de Ecclesia* VI. 1, Ed. Beck, p. 15. cited by Voobus, *op. cit.*, pp. 109-110.

27. Vööbus cites several references from the theologically unsystematic St. Ephraim's many writings and comes to the conclusion: "His (Ephrem's) definition of sin makes it impossible to accept original sin. Sin for him is a 'matter of freedom' (*Contra Haereses* XVIII. 3. ed. Beck, p. 64), and its roots are in the will" (*De Virginitate* III. 8. ed. T. Rahmani in *Hymni de virginitate*, Scharfe, 1906, p. 9). In another place, he gives a fuller definition: "the sin is this, that a nature (furnished) with will, and a being (furnished) with freedom becomes guilty." (*Contra haereses* XXI. 5. ed. Beck, p. 74). Sin cannot be located inherently in nature (*Discourse to Hypatius* V. ed. C.W. Mitchell, in *S. Ephraim's Prose Refutations of Mani, Marcion and Bardaisan*, London 1912, I. 144ff).

Therefore one cannot say that human nature has been fundamentally affected by sin or transformed into evil. (*Hymni de ecclesia* III. 4. ed. E. Beck, CSCO, Louvain 1960, p. 8).

"Since Adam's fall, outward conditions have experienced catastrophic changes, but neither human nature nor the spiritual-ethical level in human existence has been affected . . . man's freedom has not been affected. Indeed, it is intact." Voobus, *op. cit.*, p. 109.

I have not been able to check the references to see whether Voobus exaggerates or not.

the idea of natural sin. Lest this be conceived as some so-called "monophysite" heresy, let me cite the ostensibly "duophysite" Council held by "Nestorian" Catholicos Sabrisho in 596 A.D. The decree says "We are watchful and remove everyone from our fellowship who has adopted (the view which) says that sin lies in nature, and that man sins not in free will, and everyone who says that the nature of Adam originally was created immortal."[28]

If sin were a part of human nature, then we would not be responsible for what we do according to that nature, argues the whole Asian tradition including Gregory. Augustine recognizes the force of the argument, and makes what appears to be a very futile effort to refute it.

Ephraim and Gregory would argue that passion or concupiscence by itself does not constitute sin. For Augustine concupiscence constitutes the root of sin. For the Asian tradition human freedom is man's most cherished possession which makes him God-like. For Gregory, freedom of every human being is so important, that revolution will be a permanent feature of human existence, until every single human being is fully liberated from all tyranny and oppression, and have a full share in the ruling power:

> "Therefore, all human governments are subject to sudden revolutions of change, for this reason that those who are of the same honour by nature can not be deprived of their equal share in the ruling power, so long as they belong to the same (human) race, there is a natural impulse to seek equality with these who rule over them."[29]

It is fair to say that this essential element in the Eastern tradition has been largely obscured in the Western tradition, and when it comes up in the West it is usually from a heretic like Pelagius or Arminius who has not grasped the theological depths of the notion of freedom, or from a humanist like Erasmus, or again from secular

28. Chabot, Ed. *Synodicon Orientale*. pp. 198ff. Eng. Tr. Voobus, *op. cit.*, p. 118.

29. *Hai de anthrōpinai dunasteiai toutou charin agchistrophous echousi tas metabolas, hoti ou dechetai to kata tēn phusin homotimon, mē isomoirein en toē kreittoni, alla tis egkeitai phusikē, pasin epithumia pros to epikratoun exisazesthai, holtan homophulon hē: Contra Eunomium* I. 527. GNO I. 178/26-179/1. PG 45. 413 B. NPNF V:84A.

enemies of the Church. Thus the notion of freedom remains a suspect stranger to the Western theological tradition. It will so remain until a new category structure can be evolved that avoids false dichotomies like nature-grace, original sin-actual sin,[30] natural-supernatural, predestination and free will.

The point here is to insist that Gregory follows essentially the Semitic tradition, to which the Augustinian notion of "original sin" which has its roots in Gnosticism and Manicheeism, is inimical. The Semitic tradition puts the stress on human freedom and responsibility, and it is this line that Gregory also adopts.

2. Sin and Diastema

In order to understand the depth of Gregory's notion of human sin, we need to consider how he describes the consequence of sin in relation to time and *diastéma.*

One way of looking at the disastrous consequences for mankind of Adam's sin is to see how the structure of human experience has been radically altered by it. There has been, as a consequence of sin, a change in the mind and a change in the body, but

30. Even the Alexandrian tradition, which speaks of an "Adamic succession" (*diadochē*) in sin, does not make this distinction between original sin and actual sin. Didymus the Blind often spoke of the *diadochē* or succession in sin. e.g. See *Commentary on Job* (Tura-Papyrus, Teil III fol. 282, lines 20-22. *Kata diadochēn tou Adam hupodeiknus hamartian koinēn ousan anthrōpōn Didymos der Blinde,* ed. Hagedorn, Hagedorn und Koenen, Bonn 1968. For a Greek orthodox view on "original sin" see J. Romanides, *To Propatorikon Amērtēma* Athens, no date). There is still need for a thorough study of the Patristic evidence on "original sin". On the question whether Gregory himself had a notion of "original sin" much has been written. See e.g. E. MacClear *The Fall of Man and Original Sin in the Theology of Gregory of Nyssa* in *Theological Studies* 9, (1948) pp. 175-212. For a less partisan approach, and a critique of Bossuet's view that Gregory of Nyssa must have held the same ideas on original sin as Basil and the Nazianzen, see J. Gaith, *La conception de la Liberté chez Grégoire de Nysse.* pp. 111-117. Gaith refrains from denying categorically that there is a notion o f "original sin" in Gregory of Nyssa, but says: "*Mais il n'apparait pas que Grégoire admette la transmission du péché originel comme faute.*" (p. 117) The Alexandrian notion of *diadoche* in sin is reflected in Gregory's writings. See e.g. *In Inscriptiones Psalmorum.* I. VIII. GNO V. 63/15. PG 44. 480 B. But his basic view is authentically Judaeo-Christian, i.e. opposed to the Augustinian idea of "original sin". The "Gnostic" origin of the notion of original sin has been ably clarified by Johannes Knudsen in Philip Hefner, ed. *The Scope of Grace*, pp. 121ff.

what is even more significant is the change in the manner in which we experience reality itself.

Diastéma is an essential characteristic of all created existence. All creation is both *distant* from the Creator, and *extended* in time and space. But neither this distance nor this extension, which is experienced in present human existence as a spatio-temporal reality, is ultimately spatio-temporal in the sense in which we now experience it. Createdness as we now experience it, is itself a consequence of the Fall.

This is already hinted at in the treatise *On the Making of Man*, in a manner which appears to contradict Gregory's general teaching that all creation is always subject to change. Change is not, however, the whole of created existence. There is an element that remains stable and continuous.

> "For our nature is not entirely in flux and change. It would be inconceivable in general if nature had nothing permanent or stable in it at all. It would be more precise to say that in us part is stable and part is prone to alteration. For on the one hand the body changes by growth or decay, like a garment which has to be altered in accordance with the size of the person wearing it. But the form of the human remains unchanged through all this change, the nature not existing outside the distinguishing marks once for all impressed upon the human, but through all the changes of the body this distinctive identity with its manifest characteristics remains unaltered.
>
> "The change which comes about as a result of passions should be excluded from the consideration, as something external which has supervened on the true form of man. For like a strange mask the form of man is subjected to deformity by disease, which when lifted and removed by the Logos, as in the case of Naaman the Syrian, or that of others in the narratives of the Gospels, the form hidden by the disease comes back through the healing and becomes visible in its own identifying characteristics.
>
> "Therefore the God-likeness of the soul does not consist in the flux of alteration and change, but rather is attached to the

> stable and self-identical in our composite nature."[31]

Thus while change or *tropé* is the characteristic of created existence, it is not the whole of that existence. There is a continuity that belongs to the form of man, parallel to the continuing identity of a human being despite the many changes in bodily appearance which take place in him from childhood to old age.

How does the Fall of man affect this composite nature of man? Well, now the shaping of man is like a piece of sculpture being gradually brought out of a piece of solid rock. Human nature, as the internal principle of the body and soul, shapes the matter of the earth into the body and soul of man, in accordance with a pattern. As Gregory concludes the treatise *On the Making of Man*, he says:

> "We say that our all-shaping nature, taking a piece of connatural material, in itself a part of a human being (i.e. the human seed), fashions humanity. Just as (i.e. in the case of the sculptor's work) through working out the stone little by little the form appears in it, indistinct in the beginning, perfect when the work has been completed, so also in the instrument (i.e. the body) is carved out the form of the soul, which appears in proportion to the substratum in which it is manifested, imperfect in the imperfect (body), but perfect in the full-grown. If, however, the nature had not been maimed by evil, it would have been perfect from the beginning.[32] Because of this, our sharing in the impassioned and animal-like manner of generation has made it impossible for the divine image to shine forth immediately in the fashioned body, but man is led to perfection in the properties of his soul, only by a course and sequence through the material and animal creation."[33]

It is quite clear that we are in the presence of what appears to be a self-contradiction or mutual inconsistency in the thought of Gregory. For the import of this passage certainly goes contrary

31. *De hominis opificio*. xxvii. PG 44. 225 D—228 A. NPNF V. 418.
32. *All, ex archês in teleion ên, ei mê dia tês kakias he phusis ekolobōthê.*
33. *De opificio hominis*. xxx. PG 44. 253 D—256 A. NPNF V. 426B.

to the earlier affirmation that the "course and sequence" (*hodos kai akolouthia*) belong integrally to all created existence, for here it is stated that the "course and sequence" is due to the coming in of evil.

This is indeed a major contradiction and there are only two ways of looking at it. Either it is a real self-contradiction, i.e. Gregory was unaware that his two ideas were mutually contradictory and left them there because his thinking was not integral and systematic.[34] The other possibility is that Gregory was aware of this ambiguity and left it unresolved conceptually as an inherent difficulty in all conceptual expression of the nature of temporal and transtemporal reality. We tend to take the latter view, but before we deal with it, we should cite other passages where the contradiction becomes more apparent.

The clearest expression in Gregory of the view that temporal development of man is a consequence of the Fall occurs at the end of the *Commentary on the Song of Songs,* where Gregory brings in his understanding of the "double creation" of man:

> "Who will draw out for me this water lying so deep, of these ideas which are so unattainable to our reason? But it seems good to me to testify in your hearing that it is only to those to whom the Apostle says: 'Be filled with all, with all reason and all knowledge', that it is given to know these things. Our own poverty is incapable of laying hands on the treasures of the Word that are laid out before us. Rather as we would wish to remain outside the condemnation for laziness, we would not shrink back from offering something modest through the sweat of our labours, through the grace of him who ordained us to the investigation of the Scriptures.
>
> "We shall then speak of some of the more refined teachings in philosophy which through the praise of the bride is

34. This view was forcibly expressed at the Freckenhorst international conference by Prof. G.C. Stead (Ely Professor of Divinity at Cambridge), who agrees with the view of Eunomius about Gregory (see *Contra Eunomium* I. II. 592. GNO I. 339/14-16. PG 45. 1109 B. NPNF V. 310 A). Madame M. Harl, Sorbonne Professor of Classical Antiquity, who was present in the discussion did not see any real contradiction at all, and has promised to deal with this problem in her future writing. The present writer feels the problem is real, and that Mme Harl will have to justify her minimising of the difficulty.

offered to us in these words. This then is the teaching. It is not according to the same sequence and order that existents are created and then re-created. For on the one hand when in the beginning the nature of the creation came to subsist through the divine power, in each of the existents the final end was co-perfected with the beginning without an interval (*adiastatōs tēi archēi sunapērtisthē to peras*) in all that has been led forth from non-being into existence, the final perfection sprang up together with the original beginning. And human nature is also one of these created existences, so it too, like the others, did not have to progress by gradual advance to reach perfection, but rather from the very beginning of its existence was co-fashioned also its perfection. For Scripture says: Man came to be in accordance with the image of God and (His) likeness, by which is signified the highest good and the most perfect. For who can find anything surpassing the likeness to God? So then in the first creation the final end was co-manifested with the beginning without an interval, and it is from perfection that nature begins its existence.

"Afterwards, however, on the other hand, having become subject to death because of its kinship to evil, when it fell off from remaining in the good, it no longer, as in the first constitution, receive its perfection instantaneously, but advances as by a road, towards the greater through a sequence and order, little by little shedding its propensity towards the opposite. In the original fashioning there was nothing to prevent the coincidence of the perfection of the nature with its beginning, evil not being present; in the second reconstitution, however, (*epi de tēs deuteras anastoicheiōseōs*), by necessity diastematic extension accompanies the return to the good. Because our mind is now bound together with the material propensity because of evil, it is only little by little that it scrapes off and sheds as it were the mean and congenital sheath enveloping it, through a more refined conduct. Because of this we have been taught that the abiding-places in the presence of the Father are many (Jn 14:2) made ready for each one as recompense in proportion to each one's adhesion to the good and separation from the

> evil. For there are some who are still at the beginning of their coming into inheritance of their improvement, almost as if just brought to the surface, to the recovery of transformation in the truth, rescued from the deep abyss of the life of evil, transformation through attention to and step by step growth in the good, others have made greater progress through their striving after the good, some are in the middle point of their ascent on high, others have already passed the middle point, yet others have even climbed higher than these, and some others have gone even further beyond these latter, there are others who are bracing up to surpass even these in their upward climb. And wholly in accordance with the diversity of the will and purpose of each, God admits into the various orders in His presence, alloting to each according to his worthiness, leading them all together from the lesser towards the higher participation in the good. This is how we understand the difference between the souls as taught in the present passage, distinguishing the souls in terms of their vision of the Bridgeroom."[35]

One hesitates to smooth out what appears to be a basic contradiction. Gregory normally says that change is the unvarying characteristic of the created order. All creation is in change, not as a consequence of sin or of the Fall of man, but as the distinguishing characteristic of its nature as created. All creation is placed in *diastéma*, i.e. in the necessity to proceed from the *arché* to the *telos* by traversing the path between *arché* and *telos*. Here in this passage, Gregory clearly says that in the original creation there was no *diastéma* between the *arché* and the *telos*, and that the *diastéma* was introduced on account of sin and consequent attachment to material existence. It seems more reasonable to think that the contradiction is real, than to minimise it as Mme Harl did. We may not agree with the patronising judgment of Dandelot that Gregory could not avoid the contradiction because he was not a real "theologian", but only an "homme de l'eglise".[36] We have to agree, however, that Gregory has not satisfactorily resolved the contradiction.

35. *In Canticum Canticorum*. XV. GNO VI. 457/1—460/5. PG 44. 1108D—1112 A.

36. Dandelot, *op. cit.*, p. 178.

There is a definite ambivalence in Gregory who oscillates between the desire to be wholly faithful to the Christian tradition, and his passionate desire to make maximum use of the insights of pagan philosophy. He sweats profusely in his efforts to work out a reconciliation between these two approaches, and he is rightly modest in setting-forth these his most refined thoughts expressed in the last two orations of the Commentary on the Song of Songs, more in the form of a hypothesis to be examined, than as Church teaching to be accepted.

We can do justice to Gregory only by examining this apparent contradiction in greater detail, for the contradiction can lead us to a deeper understanding of Gregory's basic thought structure.

One way of reconciling the apparent contradiction is to posit a double theory of creation in Gregory. Texts from Gregory can be advanced to show that the Cappadocians held such a theory, that the original creation was of intelligible beings alone who did not need to traverse any interval between beginning and perfection, and that it was only as a consequence of sin and fall that the material creation and the coarse and heavy human bodies were prepared for human souls. Traces of this Origenic view can be seen in Gregory himself who consciously denies the pre-existence of souls as we have already shown. But it is the same Gregory who in his *de hominis opificio*, clearly admits that "that which is made in the image is one thing, and that which is now manifested in wretchedness is another".[37] He says clearly also that Scripture teaches "double creation"[38]—the original "creation" like to God, and the second "fashioning" like to the animals.

Does Gregory believe that change in space and time really belongs only to the material creation, and not to the "spiritual beings"? There are passages which could be adduced in favour of an affirmative answer to that question.[39] But then Gregory also affirms that all creation, both material and spiritual is subject to change, as we have already shown. As Cardinal Danielou has remarked, to ascribe radical mutability to the creation as a whole, and

37. *De opificio hominis*. XVI. PG 44. 181 A. NPNF V. 405 A.

38. *Oukoun diplē tis estin hē tēs phuseōs hēmōn kataskeuē, hē te pros to Theion homoiōmenē. hē te pros tēn diaphoran tautēn diēirēmenē*. *Ibid*. PG 44. 181 B. On double creation itself see Ugo Prianchi (ed.) *La Doppio Creazione dell' Uomo* Edizione dell'ateneo e Bizzarre: Roma, 1975; also Harvanek, R.F., *The Philosophy of Creation of St. Gregory of Nyssa,* (Dissert, typescript) Fordham University, 1944.

39. See *Ibid*. XIII. PG 44. 165 A. NPNF V. 399 B—400 A.

not merely to the material creation, is to take sides with the Stoics and against the Platonists. For the latter, the Spirit is unchanging and only matter is subject to change.[40]

Gregory does depart from the Stoic tradition in positing that the Creator is unchanging and only the creation is changing. And he cannot say that the creation is nothing but change, there has to be some stable element that remains unchanging, else there would be no identity for anything. This stability is confined within the boundaries of nature, for to change beyond nature is to cease to exist.

If we were to seek to reconcile the apparent contradiction in Gregory's thought, we would suggest that sin and fall have made this created changeability problematic by changing its original direction. Changeability as created is the occasion for freedom. If there were no possibility of change, there would have been no freedom to grow in the good or to choose between good and evil. But once having chosen the opposite of the good, changeability loses its original direction, i.e. towards the good, and slips into a downward flow, thus becoming the way to non-being and death.

Diastéma and mutability are aspects of creation, and as such good, for they constitute the framework within which created freedom is to be exercised. But sin makes the framework problematic by introducing a compulsive force or drive towards evil.

The body and matter in general are affected by sin. For the beauty or goodness of matter or of the material body is not in itself, but in the image of God that indwells it and lends beauty and goodness to it. So long as the mind and will of man are alienated from God, the body is a problem.

It is in this alienated existence that man lives by "memory and hope", and time is always experienced as death, every moment becoming dead as soon as it comes into being. *Diastéma* as we now experience it, in guilt and anxiety, is the consequence of sin. But *diastéma* in itself is the consequence of being created.

This reconciliation does not resolve the contradiction in Gregory's thought, but only mitigates it. For when Gregory says that in the original creation the *arché* and the *telos*, the inception and the perfection, co-existed, he is denying the very doctrine of *diastéma* which is so central to his arguments against Eunomius.

The contradiction remains, in Dandelot's happy phrase "*malaisé*

40. Danielou, *L'être et le temps*, p. 99.

a-résoudre". But it points to an *aporia* in the very nature of human thought. We cannot see time from the perspective of God to whom the past, present and future are co-present. It is only from that perspective that one could say that the inception and the perfection are co-present.

But to say that the beginning and the end are co-present is to pose the problem about the meaning of time and history.

3. Freedom and History

Existence, for Gregory, excludes evil. To exist is truly possible only in the good. There is no such thing as a "sinful existence", because what is sinful is already dead, condemned to annihilation. As he puts it most succinctly in the Commentary on the Psalms;

> "Scripture, through these words, teaches us that to exist in Him Who Is, is truly to exist. If anyone has fallen off from Him Who Is, he is no longer in existence. For to exist in evil is not truly to exist, for evil itself, in itself, has no existence, but the non-existence of the good becomes evil. Just as he who exists in Him Who Is is truly in existence, so also he who has come to be in non-being (for this is what evil is) is reduced to non-existence, as the passage says."[41]

And most of history has its origin in the movements of the will of man which is already alienated from being.

On the one hand Gregory denies that all history is merely a pre-determined drama inexorably issuing from antecedent causes. History is truly the arena in which the freedom of man operates through the faculty of choice. Here Gregory departs from all three leading schools of pagan philosophy in his day—Aristotle, the Stoics, and the Neoplatonists.[42] On the other hand Gregory rejects the Indo-Hellenic notion of Punarjanma or Metempsychosis as an explanation for history, for this too would be denying the freedom of man. Gregory boldly criticizes both Plato and Origen at this point.[43]

---

41. *In Inscriptiones Psalmorum* Pars I. cap. VIII. GNO V. 62/26—63/6. PG 480 A.

42. For a brief and illuminating resumé of Gregory's arguments see J, Gaith, *Liberté*, pp. 87-94.

43. In *On the Making of Man* and in *On the Soul and the Resurrection* one finds the most explicit criticism of both Plato and Origen on this point. PG 44. 229 B ff. and PG 46. 108 B ff. NPNF V. 419 A ff., 453 B ff.

Both these doctrines which may have originated as theodicies, that is, for the purpose of defending God's justice against the obvious injustice in the world, would make man a puppet controlled by antecedent factors over which God has no control. History is not a puppet-show. Man is free and therefore responsible.

Gregory argues against Origen's view that the soul has come into a body because of sin. But he does hold that the body has become different on account of sin. The body of man before sin is light and free-moving, while here it is beset with heaviness and inertia.

Gregory cannot, however, accept the notion which Augustine later introduced, that human nature is born in sin and can do nothing but sin. He makes it clear that this kind of "determinism by original sin" would also take away all responsibility from man, for man cannot be blamed if his nature is evil and sinful for acting in accordance with that nature. On the contrary, he could be blamed for doing any good, for that would be against his sinful nature.[44]

Man is not born from sin. He is born from the creative act of God, from whom he receives his nature. Sin is against nature. Only the good is in accordance with nature, for that nature is created good.

Despite the fall, man retains a measure of freedom. He is fallen and he is now a slave to the passions and to evil. But he is not a totally impotent slave, for if he were, he could not be held morally responsible for his actions. In history, however, God demands moral action, righteous action, from man.

But history is not the whole story. *For history itself is a consequence of the Fall.* Man, who is created in the image of God, should not live by memory and hope, since God does not live by memory and hope. This present world, in which all time is, immediately as it comes into being, swept away into the past, is a dead world. This is not our true mode of existence. We now live by memory and hope, but our true mode of existence is, like God's, without memory and hope, i.e. life in eternity:

> "For since we are not able to have all things at all times before our eyes, some of the things that are ever present we know (directly), others we register in our memories. We can-

44. *De anima et Resurrectione*. PG 46. 120 C. NPNF V. 457 A.

> not keep our memory unconfused without having the signification of names to distinguish from one another the things stored away in our minds. To God, however, all things are (always) present, and has no need of memory, his all-perceiving power dominating and knowing all things at all times. What need does he have for words or names, the wisdom and power that is in Him themselves holding together unconfused and well-distinguished the natures of all existents? Therefore on the one hand are all things existent from God and before God, on the other hand, for the sake of our own guidance, the names signifying different things exist."[45]

Here is the locus of the *aporia* and of the seeming contradiction which stems from the *aporia.* Time itself is a special kind of reality which we can neither comprehend nor transcend, since we are enclosed within it and cannot stand outside of it. But time is the stuff of history, and time is not only a created reality, but a reality that has been distorted by sin and the fall. Time as we now know it is time bound by death, on account of sin. This present time, historical time, has as end, as it had a beginning. Gregory leaves us in no doubt on that point. Individual beginning and end (i.e. birth and death) are themselves a consequence of sin. We are not born in order to die, it is because of death that birth is necessary. In "heaven" there is no death, and so no birth is necessary. For this reason our Lord said that there is no marriage in heaven either. Gregory who vigorously defends the sanctity of marriage and refuses to regard concupiscence as evil in itself, still holds that the male-female distinction does not belong to the original creation, but was added in view of sin. The sexual organs are not instruments of sin, but creations of God for perpetuating the race in view of human sin and the death that follows sin. Gregory says in the Catechetical Oration, that if the sexual organs were to be regarded as evil or the source of evil, this would be to attribute the beginning of evil to the Creator Himself:

> "The only shameful thing in itself is evil itself, and perhaps

45. *Contra Eunomium* II. 282-283. GNO I. 309/16-29. PG 45. 1005 C—D. NPNF V. 279 A. The idea that God needs no memory or hope occurs frequently in Gregory. For another instance see *de anima et resurrectione*. PG 46. 92 C—93 A.

whatever has affinity with evil. The orderly process of nature (*hē tēs phuseōs akolouthia*) ordained by God's will and law, is beyond the reach of any accusation of evil. Such accusation, made by people who regard anything connected with nature as shameful or improper, reaches up to God, the fashioner of our nature. If therefore only evil is separated from God, and nature is not evil (*phusis de kakia ouk esti*), and the mystery (of the Incarnation) says that God was born in man, and not in evil, and if there is only one way for a human being to enter into life, through which that which is to be born is led out into life, what other mode of entrance upon life would they legislate for God? (After arguing that all the organs of the body are organized for the common purpose of sustaining and continuing life, Gregory concludes:) The reproductive organs have a purpose to provide for the future, to continue the succession of the nature through counter-acting (death). But if you are considering utility, to which organs regarded as honourable, are these inferior? (See I Cor 12: 14-24) For it is neither by the eye, nor by hearing nor by speech, nor by any other organ of sense, that the continuity of our race is maintained. For these organs are, as we have said, for the enjoyment of the present existence, but it is by those (reproductive) organs that deathlessness is maintained in humanity, so that death which constantly battles against us is rendered in-effective and powerless in some way, the nature constantly filling the breeches left by the ravages of death through bringing others into birth."[46]

Yes, sex is not sin, human reproduction is not dishonourable, it is indeed more honourable, since it combats death, which is the consequence of sin.

But where there was no sin, there was no death, and so no need for birth. Where there is no need for birth, the sexual relation and marriage are no longer necessary (See Lk 20: 34-38).

History, then, is not the whole of reality. Created reality is visible in history only in its distorted form, altered by sin. Man has "lost his wings" and "the body has become heavy".

---

46. *Oratio Catechetica*. 28. Srawley 105/16—107/14. PG 45. 73. NPNF V. 497.

Gregory's view of history thus reveals again a highly sophisticated dialectic. On the one hand, history is distorted reality, where the wheat and the tares grow together, where the movements of man's free will striving for the good, and the futile movements of that same will towards evil and consequent non-being, function together, producing this strange, ambiguous, semi-real, semi-illusory, reality called history.

4. The Loss of Wings—Splashdown and Reorbiting

The idea of "wings" is central in Gregory's thought. God himself has "wings", says Gregory in his sophisticated exegesis of certain passages in the Psalms (16:8), in the Song of Songs (6:4), in Deuteronomy (32:11) and in the New Testament (Mt 23:37). The Psalmist prays for refuge under the wings of God, Moses speaks about God spreading his wings over his people. The Bride of Canticles says she has received wings. Christ, entering Jerusalem, wants to gather the people as a hen gathers her chicken under her wings. So Gregory continues:

> "If these God-inspired passages, thus ascribe wings to the divine nature in some ineffable way, and if the first fashioning of man witnesses to our nature having come to be in the image and likeness of God, it is by definition necessary that the image must bear likeness to the archetype in all ways. Now, according to the Holy Scripture, the prototype has wings, should not then the nature of man also have been fashioned in a winged manner, so that in thus being capable of flying also it should bear likeness to God? It is clear of course, that the term 'wings' has to be conceived in some transformed allegorical sense in order to be fitting to God, the term wings being applied to indicate the power, the blessedness, the incorruptibility and such like. Since these qualities had also belonged to man, so long as he was like God in everything, but afterwards when the inclination towards evil robbed us of these wings (having become outside the shelter of God's wings, we were also stripped of our wings), for this reason the Grace of God has appeared to us, enlightening us, so that setting aside ungodliness and wordly lusts, we might once again,

through holiness and righteousness, grow our wings (See Tit 2: 11-12)."[47]

The idea of wings belongs essentially to the Platonic tradition, and was introduced into the Christian tradition by Origen. According to Plato, it was a characteristic of the soul to hover around on wings (*hepterōmenē meteōroporei*) and indwell the whole cosmos (*panta ton kosmon dioikei*),[48] a kind of winged charioteer. Gregory's exegesis contains many verbal allusions to Plato's Phaedrus, and there is no need to doubt Platonic influence here. The divine presence, for Plato, is peopled with winged choirs moving about in trans-temporal space, Zeus being captain of the choirs[49] of rather divine hosts.

The loss of wings, the heaviness of the body, consequent fall into these nether regions—these are all in Plato as well as in Plotinus, who comments on the Phaedrus passage in the 8th tractate of the Fourth Ennead.[50] According to Plotinus, it is a drifting away from the universal and attachment to the particular that is the occasion for the loss of wings,[51] and if the soul immediately shifts its attention away from the particular to the universal everything will be all right.

Gregory no doubt has read all this, and develops these ideas, seeking to base them on Scripture. Dr. Scouteris has made a most useful study of the terms *meteoroporein* and *symmeteoroporein* in Gregory.[52] *Meteoroporein* signifies "flying aloft", or spontaneous upward movement, and with the prefix *sym*—, it means soaring aloft in the Divine presence, with the heavenly choirs.

Man's original "home" is in the heavenly regions, according to the whole Platonic tradition, as well as according to Philo the Jew. For Philo, as Scouteris shows, and for the Platonic tradition, it is the body, its heaviness and coarseness, that weighs

47. *In Canticum Conticorum* XV. GNO VI. 447/13-448/16. PG 44. 1100 C—1101 B. *Glory to Glory*, pp. 284-285.

48. *Phaedrus* 246 b 7ff. *Teleia men oun ousa kai hepterōmenē meteōroporei, te kai panta ton kosmon dioikei, hē de (i.e. thnēton zōion) pterorruēsasa pheretai heōs an stereou tinos antilabētai, ou katoikistheisa, sōma gēinon labousa. . .*

49. *Phaedrus*, 246 e 4.

50. *Enneads* IV. 8, 1, 2, 4 etc.

51. *Ibid.* IV. 8.4.

52. *Meteōroporein kai summeteōroporein para toē Hagioē Grēgorioe Nussēs*, in *Theologia*, Athens, 1969.

down the soul and prevents it from flying aloft. Philo regards the body, not as directly created by God, but only as an act of "tailoring" to protect him from the shame of guilt.[53] The body dominated by the irrational soul is something into which the logos or rational soul descends seeking to master it; it is a creation of God, not directly but through secondary powers.

Gregory does not follow Plato, Philo or Plotinus[54] in these views. The body and the soul are created together, and the whole is man, not just the soul. But the body as originally created, and body as it now appears, are different in character. The present human body is weighted down by evil. "God" is the fuel that powers the soul and body for its winged flight in the upper regions. The lack of good, which is evil, thus becomes a weight, an inability to take off:

> "The human nature fell away from the light and sank down towards sin, and was no longer in its right form, having been alienated from the true life itself. 'Seated' says the verse, 'in darkness and in the shadow of death, shackled in misery and iron fetters' (Ps 106:9 LXX). For it became unable to move (*akinēton*) because of the weight of the fetters gained by evil. 'Iron' or 'fetter' signifies poverty in the good *hē tou kalou ptōcheia*), like an iron casing for the hearts."[55]

The human body functions properly only when human nature is in its proper relation to God and therefore in proper form. Present history is the history of the prison-house, where the prisoners bound by shackles of iron still try to fight, love, steal, serve, grab, give and so on. This life here is not the true history of man. It is an interlude in prison—a significant interlude significant for eternity itself.

The human body in history is a distorted body. Gregory goes into raptures when he describes the body as it now is—it is still a marvellous instrument.[56] But it does not have its fullness of beauty

53. *Quaestiones in Genes.* I. 53. See Wolfson, *Philo* I. p. 269, p. 286.

54. Philo uses *meteoropolein* and *symperipolein* instead of the Platonic *meteoropolein*, but the meaning is the same—"to be in orbit". Philo in *De praemiis et poenis* (20:121) speaks about the soul becoming initiated into the divine perfections and thus being set in orbit along with the heavenly choirs.

55. *In Inscr. Psalm.* I. VIII. GNO V. 56/9-15. PG 44. 472 A.

56. *De opifieio hominis.* XXX. PG 44. 241 D. NPNF V. 423 A.

being deprived of the essential qualities of the good—namely freedom from evil, immortality and incorruptibility, plenitude of beauty and goodness, which are the attributes of God whose image man is.

Sin has distorted the body. This is what is meant by loss of wings. In its proper state the body should be free from disease and death as well as from other forms of evil. In the unfallen state of man,

> "Then, death was not, disease was absent, 'mine' and 'thine' —these bad words—were far from the life of the First Man. For as the Son is common to all and so is the air, and before all the grace of God and his blessing was common to all, so also in equal measure was participation in all good according to capacity open to all, the sickness of greed was unknown, the hatred of the superiors towards inferiors was not (in fact there was no such thing at all as superiors), and a thousand such other things which no one will ever be able to enumerate exhaustively in words, and which man possessed in his greatness, I mean equality in honour to the angels, boldness of access into the presence of God, the vision of the super-cosmic good, and all the ineffable beauty of the Blessed Nature could then be seen in us also, manifesting the divine image in themselves, in the prime of the soul when it was still shining. But those things opposed to these, the malevolent swarm of passions, the stinging wasps' nest of evil, which of these evils of this life can one say is the foremost, since they are of equal worth with each other, each of them takes the lead in prominence among evils, since each can be equally a starting-point for becoming equal to the brutes."[57]

In the passage immediately following Gregory gives an impressive catalogue of the evils of our age. But the important concern in cataloguing these evils is rhetorical in a good sense:

> "He who compares these (evils) with those (good)—shall he for ever live like the brutes? It is time therefore to take

57. *In Ecclesiasten* VI. GNO V. 386/9-387/5. PG 44. 708 C—D.

> these things into account......... having this life rectified, the grace of rejoicing through the hope that is laid in store for us is now declared. And hope maketh not ashamed—as the Apostle says (Rom 5:5)."[58]

Yes, history is the "hour", the *kairos*, for men to make decisions, whether they prefer the brute life or the life in the good that is their patrimony which is now restored to mankind in Jesus Christ. History is not the whole of truth, but an arena for self-determination for mankind. That is the challenge with which the sixth oration on Ecclesiastes concludes.

This view has important consequences. Which is the true nature of man—this sinful existence of his here and now in history, or his "original" being, which is the perfection of all good? For Gregory there can be no compromise with the tradition that develops from Augustine which holds that the nature of man is evil. Phenomenological man is distorted man, not true man. Man in history is an aberration, a straying away from the trans-historical. History is not ultimate truth, even after the Incarnation. Sin is not the final word about man. But history is real, in a qualified way, but not in itself.

The truth of history is inseparable from that which transcends history and dependent on it. History must come to an end. Gregory insists on that. Man must learn to grow wings and take off again. He does not belong to the earth. The splashdown into history is only an intermediate stop. Our place is in orbit around the presence of God. We must refuel here, get our wings fixed, and soar aloft again.

58. *Ibid.*, GNO V. 388/5-12. PG 44. 709 B.

CHAPTER VIII

THE FULLNESS OF MAN
Man Beyond History

1. Pleroma—The End of History

Pleroma, often translated 'fullness' was a technical term, whose central significance in Gnostic systems has been noted by many scholars. It was accepted terminology in the culture of the first centuries of our era, and communicated to the educated of those days a concept which has become strange to our own ears and minds. The New Testament itself uses the word *pleroma* in this technical sense without needing to give a definition of it.

For example St. Paul says in Colossians 1:19 :

> *Hoti en autoē eudokēsen pan to plērōma katoikēsai kai di, auton apokatallaxai ta panta eis auton*[1]

What is this *pleroma* that God allows to settle down in Christ as if Christ were a country to be colonized? To translate *pleroma* as fullness of God is to miss the point here totally. The writer does not say "fullness of God" (which is in Col. 2:9) but only "the whole *pleroma*"—and obviously his readers understood what he meant, for the word *pleroma* was a current term. The *pleroma* is the whole corpus of true being, i.e. God with the other "intelligible" beings around him. In Gnostic terminology, the material world is outside the *pleroma*, and the soul has to find its way back into the *pleroma* from which it has been expelled and imprisoned in a material body. For Gnostics, God is the creator of everything in the *pleroma*, but animal and material existences are outside the *pleroma*, not created by God, but by the Demiurge, an inferior and imperfect being, who

1. "For he (God) was pleased to settle the whole *plērōma* in him (Christ) and through him to reunite the universe to himself" Col 1:19.

made everything outside the *pleroma*.[2] The *pleroma* itself in the universal terminology of the Middle East in the first centuries of our era refers to God and the 'spiritual beings' or 'aeons' that are in the true world.

In the Ptolemaic doctrine of *soter*, for example, the distinction is made between *pleroma* and *eikones* i.e. truth and shadows. Everything that comes from the syzygies (yoked pairs) of aeons belongs to the *pleroma*, everything coming from single existences comes under the category of eikones or images.[3]

But *pleroma* has also an ordinary sense—as full number, crew or as complement needed to make full. It is this latter non-technical sense that dominates Gregory's thought, but he almost makes it into a technical term. He can speak of the *pleroma* of angels, meaning the whole company of heavenly beings.[4] God is the fullness (*pleroma*) of all good;[5] the fullness of all the elements is present together in beings.[6] The only Begotten Son has the fullness of Good and the fullness of Being in himself[7] as Gregory never tires of repeating.

Our interest in the term *pleroma* here is limited to the idea of the "full complement of humanity". This idea is a key to Gregory's understanding of history and of the Incarnation and even of meta-history. The whole of humanity, in all places and all times, together constitutes the human nature in which Christ has become incarnate and which Christ has united to himself:

"For all of us together are the human nature, this is the sheep

2. See Ireneus *Adversus Haereses* Bk. 1. Chh. 1-5. This refers only to Valentinian Gnosticism, but it is probable that the word *pleroma* had a common meaning both in Gnostic and non-Gnostic writings—the full corpus of true beings, as distinct from the imperfect material creation which has only partial being.

3. *Hosa ek suzugias proerchetai plērōma estin, hosa de apo henos, eikones* in Clement of Alexandria, *Stromateis* 4:13. Pg. 8:1297B. Cited in GWH Lamps, Ed. A *Patristic Lexicon*. S. V.

4. *Hapan tōn aggelōn to plērōma* CE III:48 (GNO II:68/10-11; PG 45:646 A).

5. *Pantos agathou plērōma*—GNO II:20/25. Pg. 45:581 C; GNO II: 151/29-30; PG 45:729 B Etc.

6. *de anima et Resurrectione*. PG 46:85A

7. CE III:1:81. GNO II:32/9-10 PG 45:593 C; GNO II: 188/10 PG 45:772 A etc.

> which the Good Shepherd has saved again by becoming the First-born (of the new humanity)[8]

Gregory makes it even clearer in *On the Making of Man* but he prefaces his remark with a qualifying introduction, reminding us of Origen's distinction between doctrine and speculation. He knows the thing is difficult to state verbally or even grasp conceptually. In so far as Man is the Image of God, he should be like God—immortal perfect, changeless. But the origin of Man is different from that of God. God has his being in himself and is in no need of becoming. Man on the other hand comes to be by becoming, i.e. coming into being out of non-being. He is subject to change, subject to the possibility of mis-using his freedom and becoming corruptible and mortal.

God who knows the end from the beginning, both end and beginning being co-present before him, also 'provided' the male-female distinction and sexuality as a means of human nature surviving the death of the first Man, a device which makes man kindred to the animal world.

It is as a hypothesis for discussion that Gregory speaks of two creations—the first one in which human nature as such, as a single entity, was created all at once. All created entities are finite. Humanity or the human race is also a finite entity, and it does not go on multiplying indefinitely. The curtain must fall some day on this present stage of human existence with birth and death and multiplication. The *pleroma*, the fullness of humanity is thus a limited entity with a beginning and an end, and when that limit is reached history itself as we now know it must come to a close. That finitude is an essential character of all created realities including humanity as such and time as such.

It is this whole *pleroma* which was created at first. It is this *pleroma* that Christ has leavened, like leaven hid in three measures of flour. It is this whole *pleroma* and not merely individual man that is in the image of God.[9] We have, as the human race, a double fashioning (*diplē tis*) Gregory proposes. One is a nature like unto God, the *pleroma* of humanity being the *pleroma* of all good, which

8. *Contra Eunom* III:II:49. GNO II:68/20-22; PG 45:636 B.
9. *de opificio hominis :* XVI PG 44:185 C NPNF V: 406.

in fact is God's nature. This is what was the original creation. The other is the individuated human nature, divided into male and female. This distinction of male and female does not belong to the original creation nor will it survive in the final fulfilment. In Christ there is neither male nor female, says the Apostle.[11]

But this dual nature, the one like unto God, and the other like the rest of creation—animals and vegetative life and inorganic matter is what becomes the ground of our mediating role between Creation and Creator. Both creations are then real. It cannot be said that creation in our kind of time, with its birth and death and individuality, has no function. Gregory would not accept the *Advaita* doctrine which denies all significance to historical existence. It is through this being played out in history that the *pleroma* of humanity draws the whole creation into itself and becomes enabled to fulfill its mediating role between Creator and Creation. By being born like animals and being nurtured with material food, the *pleroma* of mankind in history assimilates the rest of creation into itself and becomes integrally related to it, so that the redemption of the *pleroma* of mankind is also the redemption of the whole of creation.

Time itself is thus a time for shaping and completing the elementary form of humanity. History is the nursery and *Kindergarten* of Mankind. There is a certain significant element of the shaping of the whole of mankind and of each individual here in history, but this is only a preparation for the true life of the *pleroma* of humanity that will begin only with the re-constitution of the universe beyond time.

As Gregory never tires of saying, winter is past, Christ has come, we are in the spring; but beyond time itself lies the glorious summer, when the crops will be gathered and the feast made ready.[12] Mankind, during the hard winter of the reign of sin, darkness and idolatry, became petrified, like the very idols man worshipped.

> "Turned thus to stone by the worship of idols, human nature became immovable and unable to advance: it had become

10. *Ibid.* PG 44:184 B—"If the divine is the fullness of all good, then this (human nature) is the image of the other".

11. Gal 3:28.

12. *de opificio hominis* XXII. PG 44:208 D-29 A NPNF V:413.

> stiff with the chill of idolatry. And therefore the Sun of Justice rose in the cruel winter; the spring came; the south wind dispelled the chill, and together with the rising of the sun's rays warmed everything that lay in its path. Thus mankind, that had been chilled into stone, may become warm again through Spirit, and receiving heat from the rays of the Word, may again become as 'water leaping up unto eternal life.'[13]

Present history then is the spring-time of the Spirit, the time for repentance, for mankind as a whole to turn to righteousness and the good, before the summer comes when time shall cease and all things shall be dissolved and reconstituted.

History is not final reality. It is a time of freedom, when the fashioning of man is completed, by the very choices that individuals, groups, nations, and now-a-days, humanity as a whole, can make and execute. The *pleroma* of mankind and the *pleroma* of time are co-terminous. Each has a perimeter, a *terminus ad quem* after which all this business of birth and death, fights among individuals and groups, division among churches and nations, the greed and selfishness of man, the violence and aggression of societies,—in short the stuff of which history is constituted—, must come to an end.

We have to move on into a new kind of time—the *aion*, where the *pleroma* of mankind, the whole of humanity in time and space, will be co-present. That is the main difference between *chronos* (time) and *aiōn* (created eternity). In *time* Man appears only as passing away, each moment being merely a point of inter-section between the future and the past. In *aiōn* or redeemed time, the *plērōma* of mankind will be present, there being neither birth nor death, past nor future, all being always present. Mankind would shed its 'coat of skin', the material body which is the locus of individuality and death, of the attachment to the sensible and of the dominance of the passions. The coats of skin, the male-female distinction, the passions that rage in the body, birth and death—these are strangers to our true nature.

Shedding the 'coats of skin' does not mean that man will become a disembodied spirit. Only the coarseness and heaviness of the body has to be eliminated. The reconstitution at the end, when

13. *In Canticum Canticorum:* V GNO VI: 147/15-148/1 PG 44:865 D Eng: Tr. *From Glory to Glory* pp 184-185.

time and humanity will have reached its *pleroma*, will provide human beings with new bodies. The passions and desires which now disturb us will become themselves transformed into the striving after the good, the love of God.

Life in history is but a seed of corn. It must die and be reborn. One need not look for the form and shape of the full-grown plant or its functions in the seed itself. The seed of man does not look like a man. But all the possibility of Man is contained in it. History is like that. As Gregory makes his sister Macrina say in the treatise *On the Soul and the Resurrection:*

> "It does not surprise us that in this diminutive offshoot of man (the human seed) we cannot see all the creative energy and movement of the human soul; for neither in the seed of corn do we see immediately the manifestation of the full-grown plant. For how indeed can something so (large) be contained in something so (small)? But the earth keeps feeding it with nutrients taken from its own elements, and the grain becomes the plant, without changing away its own nature in the process but rather only affirming it and developing it into full growth through the energy provided by the food. As then in the case of seeds planted growth to perfection takes place gradually (*katā logon*), in the same way also in the human constitution, the power of the soul is manifested in proportion to the size of the body. Thus first in the foetus is manifested the powers of nourishment and growth. Afterwards it comes out into the light, for the sake of exercising its powers of the senses; then later, as the plant has grown, so to speak, as if a fruit borne on it, the power of reasoning is manifested in limited measures, not all at once, but developing along the whole course of the plant's growth...But that there is a necessary stopping to the growth of the number of souls, our reason can clearly see. Nature is not an ever-flowing stream, pouring out generations after generations in eternal succession and never-abating flow. It is necessary then that our nature should also have a stopping-point to its multiplication, for the following reason, we say. Every intelligible nature is constituted within its own pleroma; so it is reasonable to assume that humanity will also arrive at its limit, for in this it can not be different from other intelligible natures; it cannot always appear unfinished and

> defective. A perpetual process of addition of new generations would amount to an accusation of deficiency in our nature. So when humanity would have attained its own full complement (*plērōma*), then shall also stop altogether the flux and movement of this nature; having reached its nccessary goal; and another different constitution of life shall follow the present one with its procession of birth and dissolution. Birth having ceased, by necessity dissolution will also end. The possibility of decomposition exists only in that which is composed—and birth is a sort of composition. Necessarily therefore if this synthetic composition does not precede, obviously no decomposition can follow. So then it seems logical to assume that the life after will stand as indestructible, unaltered by birth and death."[14]

Man in the world of history is like a seed in the ground, an embryo in the womb. His full potential is not at all evident here. No phenomenological analysis of the grain of corn can reveal the nature of the maize-plant. No chemical analysis of the human seed can manifest the nature of the full-grown man. So also no amount of biology, psychology, sociology and history, can reveal to us the true nature of man.

The most important aspect of the nature of Man—namely that the whole of mankind as a single unit is to be redeemed from sin, mortality and individuality, and that reunited in Christ the whole of mankind is to bear the image of God—is beyond the reach of our secular anthropologies. It is a precious revelation of the Christian tradition, to which we can respond positively or negatively.

The whole *pleroma* of mankind is the lost sheep which Christ came to seek and to save.[15] Christ is Himself the *pleroma* of all Good; there is no Good that is outside of him. Gregory repeats again and again that favourite expression of his in relation to Christ—*pantos agathou plērōma*—the fullness of all good.[16] In him the fullness of God and His Goodness dwells bodily, and He, the Incarnate Christ is thus the Image par excellence. But Christ is also the

14. *de anima et resurrectione* PG 46:125 C-126 C NPNF V:459.

15. GNO II: 68/20; 69/10 ff; GNO V: 305/2 ff.

16. To cite but a few texts: GNO II:20/8 ff; II:151/30 II:188/10. II:222/3-4; II:254/15 ff; III:1:7/1. V:58/5 ff. V:87/17.

aparche tou koinou phuramatos[17], the first-fruits of the common lump of the new humanity. Christ is the First-Born (*Prototokos*) of the new humanity, and in him the whole human nature has been recreated.[18] He has shared our flesh and blood, and we shall share his new humanity. The whole human nature has now been exalted to the right hand of God, says Gregory in his anti-Apolinarian treatise:

> "Scripture says : 'God has exalted him above...it is clear that he who is the exalted one needs not to be exalted; it is the lowly that is exalted above, now becoming what it was not before. United to the Lord, the human nature itself is now exalted above, along with the divinity, and that which is exalted above from lowliness is this human nature. The lowly nature is the form of the servant which has now become Christ and also Lord, through the exaltation. And the Man Christ (*hō kata Christon anthrōpos*), was given his own proper name, following human custom, and in accordance with the annunciation to the Virgin, the human nature, as the scripture says, was called Jesus, the divine nature being un-circumscribable by any name; but the two have become one by commingling".[19]

The exaltation of human nature to the right hand of God thus determines its true locale—not in history, but in meta-history. History is the womb in which the embryo is formed, but it is being prepared to go out of the womb, through the trauma of death into meta-history. The First-Born, the *Prototokos*, Christ, is that precisely because he was the first to be born this way through death and resurrection.

> "For as the First-Born from the dead he becomes the first to loose through himself the pains of death so that he may open for all this way of birth through the resurrection".[20]

17. Col. 2:9, Romans 11:16. In *Canticum Canticorum:* XIII GNO VI:391/5 PG:44 1056.
18. CE III:II:50-57 GNO II:68/26—71—21. PG 45:636 C-640 A.
19. *Antirrheticus adv: Apolinarium* GNO III:I:161/5-18 PG 45:1165-C-D.
20. CE III:II; 50. GNO II:69/1-3 PG 45:636 C NPNF V:158 A.

or a little later:

> "We have known a double creation of our nature. In the first we were fashioned; in the second we are refashioned. But there would have been no need of this second creation, if the first had not been rendered useless by our disobedience. This having aged and become ready to disappear, it was necessary to have a new creation in Christ, as the Apostle says... (Col. 3 : 9. Eph. 4 : 24; 2Cor 5 : 17). It is the same One who is the maker of human nature, in the first as well as in the second. Then, he took the dust of the earth and fashioned man; now, again he takes the dust from the Virgin, and does not simply fashion a man but fashions him around himself. Then, he created; now he is created. Then the Word made flesh; now, the Word became flesh, so that our flesh may be trans-fashioned into the Spirit through his participation in our flesh and blood. Of this new creation in Christ, which he himself inaugurated, he is the first-born, having become the first-fruit of all those who are begotten into life (baptized) and quickened again by the resurrection from the dead, that he might reign over the quick and the dead, and that he might by himself as first-fruit render the whoie lump holy along with himself."[21]

It is this whole lump—*holon to phurama*, that is the *pleroma* of mankind. Not just everybody who is born by the method of generation provided in the first creation, but those who are born into the new humanity. It would be a great misunderstanding of Gregory's thought to assimilate it into our bland, tolerant, liberal universalism, which seems to believe that 'somehow' everybody will be saved.

Is Gregory then of the view that only Christians will be saved, and that others will 'go to hell'? He does not directly deal with the question. But in many places he has left indications of how he would answer this question.

2. Being and the Good—Meaning of History

If History is only a passing show, does it have any meaning at all? We have compared History to a womb but the compari-

21. *Ibid*: 52:54 GNO II:69/22-70/18 PG 45:636 D-637 B NPNF. V:159.

son does not come from Gregory and is certainly overdrawn. It is more than a womb, for in the womb the embryo has little possibility to take any decisive action about its own nature and being. But History is precisely the arena in which man can become a co-creator of himself, so that in this too he becomes truly the image of God.

The words that Gregory frequently uses, apatheia and *arete*, translated as dispassionateness and virtue, are grossly open to misunderstanding. *Apatheia* is not the absence of sin. Both are active functions of the created free-will of Man. Gaith rightly criticizes Danielou's inadequate analysis of these two terms. The latter properly places the two terms in their Platonic context, which makes *apatheia* the goal of *arete*, and provides *katharsis* or purification as the way to active apatheia.[22]

Danielou exaggerates, however, when he insists that the three terms *aretē*, *apatheia* and *katharotēs* have exactly the same sense in Plato, Plotinus and Gregory. He is right in regarding, along with Festugiere, these terms as of nonbiblical origin, developing their own centrality in Christian thought, a centrality characteristic more of Plotinus than of the New Testament, But Gregory is the inheritor and developer of a Tradition of Christian thought in which these terms played a central role. Festugiere cites Clement of Alexandria, Origen, Nyssa, Evagrius and Pseudo-Dionysius as the architects of this para-evangelical tradition.[23]

In the Stoic tradition, which deeply influenced the Christian tradition, *apatheia* denotes the entire absence of all desire and passion, as in the Hindu tradition. In the Platonic tradition, especially in Plotinus, one sees the same goal—the absence of all passion and desire. But Passion for the Stoics and Neo-platonists belongs essentially to the body. Plotinus is every clear on this point:

> "As the soul is evil by being interfused with the body and by coming to share the body's states and to think the body's thoughts, so it would be good, it would be possessed of virtue, if it threw off the body's moods and devoted itself to

22. Danielou: *Platonisme et Theologie Mystique*. p. 92.

23. *L'enfant d'Agrigente*. Paris, 1942 p. 149 cited by Danielou, *op, cit.* p. 92.

> its own Act—the state of Intellection and Wisdom—never allowed the passions of the body to affect it—the virtue of *sophrosynē*—knew no fear at the parting from the body—the virtue of Fortitude—and if reason and the Intellectual-Principle ruled without opposition—in which state is Righteousness. Such a disposition in the Soul, become thus intellective and immune to passion, it would not be wrong to call Likeness to God; for the Divine, too, is pure and the Divine-Act is such that Likeness to it is Wisdom.[24]

We can easily see both the similarities and the dissimilarities between Plotinus and Gregory. The words are all there in both—*apatheia, arete, dikaiosune, homoiosis pros theon*—but do they have the same meaning in both? For Plotinus the body is evil and the source of evil, and the trouble with the soul is that it cannot act alone without the body. It becomes free when it is liberated from the body and from its acts and desires. Righteousness is an act of the soul without the body. For Gregory who believes that Man was created as an integral body-soul organism, freedom does not mean the absence of the body and its acts; rather, it is the right use of the material body and its feelings and desires that becomes the basis for virtue.

For Gregory, unlike Origen and Plotinus the disciples of the Indo-Hellenist Ammonius Saccas, the body as such is not the "coats of skin" which God gave to man as a consequence of his sin. It is rather animality and mortality. The pattern of birth and death that is common to the animal world—that is the consequence of Sin. The words he puts into the mouth of his sister Makrina are significant in this respect:

> "Just as a man, clad in a ragged coat, when divested of it, does not any longer feel the disgrace of that which has been cast aside, so also when we have cast off this mortal and odious garb, which leathery coat has been put upon us from the animals' skin (by skin we should understand, it seems to me the form of brute nature—*to schēma tēs alogou phuseōs* with which we were clothed when we became acquainted with passion), all that belongs to that brute skin we shall

24. *Enneads* 1:2:3. Stephen Mckenna's translation p. 32.

> cast away from us along with our casting off that coat of skin around us. And such accretions to our nature through the brute skin are the mixing (of the human seed), the conception (or composition), the parturition, filth, suckling, feeding, excretions, the graduality of growth to full stature, prime of life, old age, disease, death."[25]

But for Gregory, as for St. Paul, to be divested of the 'coat of skin' does not mean that the soul exists without a body. A new coat, which is actually our 'original' coat is to be restored to us—, the garment of incorruption (*chitōn tēs aphtharsias*), which is the resurrection body. The tissue and texture of this resurrection body of ours is in part determined by our own choices in the world of history. We can 'choose our own parents' in this life of history, as Gregory says in the Great Catechism:

> "all other things that are born come into existence by the impulse of their progenitors, but spiritual birth is prepared by the power of him who is born. Herein is the danger, that he may misjudge what is advantageous, the choice lying before everyone's authority: it would be well, I say, to foredetermine by his own consideration, his impulse regarding his own birth, to choose what kind of father is beneficial, and from what it is better for him to have his nature constituted. For as we have said, it is within the power of the child in this case to choose his parents (*kat' exousian tous gennētoras ho toioutos aireitai tokos*). All beings are divided into two kinds, i.e. created and uncreated, the uncreated nature has the property of being not subject to change and alternation in itself, the created is subject to change by alteration; so it is up to the one who is to be born to choose through calm deliberation, of which of these it is for his benefit to be born as a child, of that which is understood to be ever-changing, or of that which possesses a nature which is unchanging, steadfast, ever remaining the same in the Good...For that which is born is by necessity akin to the parents. Which then is more advantageous—to enter into the unchanging life or again to be tossed about in the life of instability and alienation? It is thus clear to everyone who shares

25. *de anima et resurrectione* PG 46:148 C-149 A NPNF V: 464-465.

> in however small a measure in understanding, that that which stands is much more valuable than that which is unstable, and the perfect much more than the deficient, that which has no need, than that which is in need, that which has nothing more to advance towards and ever remains in the perfection of the good than that which is ever in need of advancing. It is then absolutely necessary for all endowed with mind to choose one of these two—either to believe the Holy Trinity as Uncreated in Nature, and to make him the Ruler of one's own life through the spiritual birth, or to regard the Son and the Holy Spirit to be outside of the First, the True, Good God's Nature, i.e. of the Father".[26]

The passage is beset with problems of interpretation and suspicions of contradiction. We have to keep in mind that the catechetical oration seeks to set forth the teaching of the Church for ordinary intelligent Christians and is not to be looked upon as a philosophical treatise. But this does not fully mitigate what appears to be a contradiction with his assertion elsewhere that the created nature can never be stable and unchanging. Change and movement are in some cases asserted to be the essential characteristics of created nature; but here in this passage in the Catechetical Oration, we are asked to choose between the unchanging and the changing. In other places Gregory asserts that perfection itself for the created nature means perpetual advance in the good. Here we are asked to choose between what is always in need of further advance and that which has nowhere else to advance towards.

But the main point here is something else—the capacity of man to determine, by free choice, the essential nature of his being, i.e. to be a co-creator of himself along with God. This Gregory regards as an essential aspect of Man's being in the image of God. As God is Creator, Man has also to become a creator, a creator of the good, a co-creator of Man.

The basic distinction between the Creator and the Creation remains—namely that the former has His being in Himself, not derived from or contingent upon another, not having come to be, neither becoming nor going out of existence; while the latter, i.e.

26. *Oratio Catechetica*: XXXIX Srawley pp 154/3-157/6 PG 45:97 D—100 C NPNF V: 506-507.

the created existence, has its being derived from and contingent upon the Creator, and having come out of non-being into existence, is always either becoming or going out of existence.

But Man in creation is given a special privilege of transcending his createdness by his freedom to co-create himself—not apart from God or independently of Him, but precisely by turning towards Him in repentance, by being united with Him in baptism and by continuing to be ruled by Him through the life of *arete*.

And in so far as he transcends his createdness to share in the nature and being of the Creator, he overcomes the *diastēma* itself, which is an integral element in his created existence. In proportion to his growth in the good, Man becomes more and more like God. And for God, the creation is always present without any *diastēma* either as gap or as extension, without past or future without before or after, the end being co-present with the beginning. Does Gregory want to say that as Man grows more and more into the image of God, he gradually sheds his very diastematic mode of existence and changeability and becomes like God, without *diastēma* or change? If he did want to say that, he was never lacking in an adequate supply of words to say precisely what he wanted to say. But since he has not clearly said that, we can only guess that he was vaguely aware of this possibility, but refrained from giving expression to it for fear of appearing to go far beyond what was clearly taught in the church.

True being then, is the being of the Creator, and we created human beings can share in true being only by conformity to the Being of God through participation in God's Being and Life.

Now exactly does this participation take place? Gregory is very traditional in his answer—by Repentance, by Baptism, and by the Life of *aretē*. Let us make clear that this is his basic structure of soteriology—not a so-called 'mystical experience' as some authors who have written on Gregory seem to leave the impression. What is called the 'mystical vision of God' is only an aspect of the turning towards God in repentance, being co-crucified and co-resurrected with Christ, being united with Christ through Baptism, Chrismation and the Eucharist, and the life of *aretē* lived out in history. It is interesting and important to note that he says nothing about the 'mystical ascent' in the *Great Catechism* or in the *Treatise on the Soul and the Resurrection.*

The ideas of 'mystical' ascent and vision of God do play a central role in both the *Life of Moses* and in the *Commentary on Canticles*. Are these two different soteriologies in Gregory's thought unrelated to each other? Are they to be reconciled in some way? The question is important for understanding Gregory's thought and has to be dealt with. But it would be a fatal misunderstanding of Gregory's thought to separate too easily his concept of the vision of God from his understanding of the 'mysteries'—i.e. the 'sacraments' as they are unfortunately called in western languages. It is a sacramental mysticism, a mysticism based on Baptism, Chrismation and the Eucharist—a point not adequately stressed by either Danielou or by Völker who have essayed to study Gregory's 'mystique' by looking at it from the perspective of later developments in European mysticism.

The meaning of history then is not to be sought in history as a self-contained realm. Human historical existence derives its meaning from its two transcendent poles—man's origin in God and his destiny in God. In itself, the world of time and history can have no meaning at all. For it is neither self-generated nor self-directed. It is an arena where important decisions are made; the importance of these decisions cannot be manifested entirely in history. What gives meaning to history, in Gregory's view, is the recognition that all historical existence is derived existence, i.e. created being, which has now been given, in Christ, the possibility of being partly a co-creator of itself, through the free choices made in history.

What is the nature of historical choice? What are the options available to man? Does he choose and effect his decisions entirely on his own? What about the Grace of God? What is the relation between the Grace of God and the effort of man?

3. The Life of Virtue and the Effort of Man

The idea that man's resemblance to God was primarily in a life of virtue, was by no means original with Gregory of Nyssa. It is at least as old as Plato. Aristotle and the Stoics never fundamentally departed from this notion, so much so that this was one of those few areas on which there was basic agreement between the three majors schools of Greek Philosophy.

Plato had set forth his ideas on virtue in practically all of his works, though special mention should be made of *Timaeus, Prota-*

goras, *Phaedo* and *The Republic*. Phaedo, which is regarded by scholars as Plato's faithful reporting on the ideas of his mentor Socrates, lays down the principle of 'imitation of God' as the true virtue of the soul.

Aristotle, the genius of a secular vision of reality, defined virtue "as a disposition of the soul in which, when it has to choose among actions and feelings, it observes the mean relative to us, this being determined by such a rule or principle as would take shape in the mind of a man of sense or practical wisdom"[27].

Aristotle was clearly aware of the defects of his definition of virtue as the "mean" between two extremes. Obviously virtue could not be the middle point between good and evil, but, as he insists at the end of Book Two, Chapter Six, virtue seems in fact more of an extreme rather than a middle point, when looked at from the perspective of a scale of good: virtue is, after all, the most right and best. The middle point in a scale of evil certainly cannot be virtue.

In Gregory of Nyssa's day, Stoicism still had its pervasive influence both among Christians and pagans. It is well known that "ethics" constituted the most important element of Stoic teaching. Already with the early Eleatic School, the principle had been accepted that the only true good was "virtue" and the only real evil "vice". Both virtue and vice were dispositions of the will; a certain *hexis* or way of holding things together, which determined the fundamental nature of persons as well as objects. In this the Stoics had simply adopted Aristotle's view that virtue was a *hexis proairetikē*—a disposition of the faculty of choice[28], not of feeling, or of capability, but related to the way in which our ruling faculty holds the feelings and capacities in control and directs them to properly chosen ends.

Nyssa simply acknowledges this basic Platonic-Aristotelian-Stoic definition of virtue, and asks, as we have already noted, in his Antirrheticus against Apolinarius :

> "Who does not know that virtue is the right direction of the faculty of choice? And the flesh is the instrument of choice,

27. *Nicomachean Ethics* BK II Ch. VI. Eng. Tr. J.A.K. Thomson, *The Ethics of Aristotle*, Penguin, 1970, p. 66.

28. See *Nicomachean Ethics* B. Two: Ch. V (105 a30).

> pressing forward by the effort of the understanding, being transformed into that towards which the motive force leads it. The faculty of choice is thus nothing else than the mind and the disposition towards something".[29]

Any doctrine of sovereign grace would be totally incompatible with such a doctrine. Augustine was perhaps fortunate in the limited knowledge of Hellenic philosophy he possessed. No one who had been initiated into Hellenic thought could say, as Augustine does in the second chapter of *de Correptione et Gratia*, that the Grace of God does not suppress human collaboration. That seems like a concession reluctantly granted to the efficacy and indispensability of the human will. For a mind like Gregory's, trained in Greek philosophy, the very idea of undermining human freedom is tantamount to blasphemy. That freedom is his most precious gift, and if the grace of God is an external cause making man do good by necessity, then human freedom has disappeared and man is no longer divine or human, but only like plants and animals, the slaves of necessity.

One could sympathize with Augustine for his failure in grasping this central insight of Eastern patristic thought, for Pelagius was no authentic interpreter of what he had learned in Jerusalem and Asia. Pelagius or his disciples had reported the teaching in Jerusalem as insisting that "the Grace of God is in accordance with our merit"[30]. Augustine insists that this teaching of the bishop of Jerusalem is "alien to catholic doctrine...inimical to the Grace of Christ"[31]. If the Grace of God is "in accordance with our merits", then "grace is no longer grace.[32] Grace, by definition, is "unmerited gift" and it would be a total contradiction to say that this unmerited gift is according to our merit.

Has Pelagius reported adequately and accurately what he heard in Jerusalem? Was the bishop of Jerusalem innovating? Was he

---

29, Antirheticus Adversus Apolinarium GNO III:I: 198/1-5 PG 45:1217 C.

30. "...dicunt, gratiam dei secundum merita nostra dari"...*de gratia et libero arbitrio*: 10 (Ch. V) Augustine sees the origin of this teaching in the Orient..."in Oriente, hoc est, in provincia Palaestina, in qua est civitas Ierusalem", where Pelagius heard it from the bishop himself.

31. *de gratia et libero arbitrio*: 10 (cap. V.).

32. *gratiam non esse gratiam*: ibid: II.

teaching something "alien to Catholic doctrine"? Unfortunately we cannot answer any of these questions, What we can do is to pit the teaching of Augustine directly against the teaching of Gregory of Nyssa, who seems to have been the source of the teaching of the bishop of Jerusalem.

It would appear at first sight, that what Pelagius reports is the true teaching of Gregory. Warner Jaeger's *Two Rediscovered Works* notes that it is a key concept in Gregory's thought that God's grace increases as Man's effort intensifies.[33] Jaeger contends that the notion of *synergia* is central and essential for the system of Gregory. He also contends that Gregory is a sort of anti-Barthian, who thinks that man has to take the initiative and that God cooperates with man to bring the effort to fruition.

The centrality of the notion of *Synergia* in Gregory's thought has been demonstrated by Jaeger only with reference to one of Gregory's disputed works—*de Instituto Christiano.* Its comparatively rare occurrence in the other writings of Gregory should at least be considered as one reason for not too easily accepting the *de instituto Christiano* as authentically Gregorian.

Jaeger also draws our attention to the distinction drawn by Clement of Alexandria[35] between *sunergeia* and *sunaition*, the latter being a Platonic concept. A *synaition* or co-cause is able to produce an effect in cooperation with something else, which by itself cannot produce that effect. According to Jaeger, a *synergon* is something which in itself cannot produce a given effect, but which, when added to something that *is able* to produce the effect by itself, makes the effect better and stronger. Clement was basically of the same view as Augustine,[36] that our cooperation (*synergia*) is necessary for the grace of God to be effective. However, Augustine only reluctantly conceded that human assent and cooperation were necessary for the Grace of God to operate effectively. Clement, at home in Hellenic thought, gladly conceded the need for human effort as a collaborator with God's grace. God is the teacher who gives the grace and knowledge. The Christian is a cooperative

33. see pp 86-96, esp. p. 87.

34. The term occurs at least 24 times in Jager's edition of *de Instituto Christiano* 44:16, 46:4, 7, 25, 26; 49:8, 54:11:55:14; 57:13; 58:14; 61;19; 70:11; 71: 23; 77; 78:13; 84:1. 9; 585:9, 15, 20, 86:4, 87:3, 19; 88:2.

35. Jaeger *op cit.* p. 104. Stahlin, Clemens Alexandrinus III:36, 10-11.

36. *de correptione et gratia Cap.* II.

and obedient student. The initiative is with God. Man responds to it in obedience and cooperation. Man's *synergia* makes God's *ergon* efficacious in his life. By himself God does not do acts of virtue but only through the *synergia* of Man.

As Jaeger points out, there is the other view also in Clement, namely that as man grows in grace, and thus becomes worthy of God's greater grace, God helps him in his initiative to practice virtue[37]. For this help rendered by God to man, Clement does not use the term *synergia*, but the verb which means to assist or to come to one's aid.

Gregory of Nyssa, however, insists on the human initiative, much more even than Clement, as a matter of principle. He sees no contradiction between divine grace and human effort. For Gregory, the human initiative is essential, for without it there is no freedom; without freedom there is no moral good; and without moral good there is no Image of God. While Gregory can readily admit that no man is good[38], he does not make that into a theological principle for arguing that nothing good can come out of man. Gregory is more realistic than Augustine at this point. He sees the dominance of evil in the majority of men; but he has to take into account the opposite fact also—there *are* a few rare human beings who are good in a large measure, in the history of mankind. This latter fact, for Gregory, is evidence that man is capable of good. If one looks at man phenomenologically, one finds that man is capable of both good and evil

> "Therefore, they are in some sense excusable, who cannot, when they look at such cases (of evil men), readily admit that the divine form is in such people. However, through those who have rightly ordered their lives, we can see the divine image in man. For if some one who is carnal and a slave of passion makes it unbelievable that man was originally adorned with divine beauty, other persons who practice noble virtue and keep themselves pure from pollution should confirm you in the better conception of human nature".[39]

37. Stahlin, *op. cit.* III: 36:1-8 see Jaeger, *op. cit.* pp 104 ff.
38. PG 45:1297 A.
39. De hominis opificio: XVIII PG 44:193 D NPNF V:408 B.

For a hundred Jechoniahs, the old Testament produces at least one Moses. And it is in Moses, rather than in Jechoniah that we should look for the image of God. And if one says that no good can come out of Man it must apply to people like Moses, Elijah and John the Baptist.

All is grace. God alone is good. God is the source of all good. Man's goodness is not self-derived, but given by God. But it is there, even after the Fall. Man is not all evil[40], though no man is all good. Human nature cannot be defined as wholly evil. It is still capable of good and evil; the capacity for good as the original gift of god is still there, though obscured and hindered by our animal propensities. No doubt Gregory is profoundly influenced by the Hellenic tradition at this point; for the Greeks purity of the mind was the way to the Divine. Gregory's doctrine of "virginity", which later became so pervasively influential in the monastic tradition, owes much to this Greek notion of purifying the soul from its propensity to animality and evil[41], through the life of virtue.[42]

The effort of man in no way denies the grace of God. The very capacity to make an effort towards the good is a gift of grace. Everything that happens for the salvation of man and for the good of the human race, comes from God. Gregory leaves us in no doubt on this matter. To take just one of many passages about the grace of God in Gregory:

> "Whatever has come to be from God for the good and the salvation of Man, it has all happened because of (God's) grace and goodness, since we do not have in ourselves the originating cause of the Good (*pros to agathon aitian*); on the contrary, since we have come to be in all kinds of evil,—evil itself having no natural existence of its own, what it is,

40. *To kakon toinun ekleipein euchetai hode anthrōpos kakon ouk estin. Pōs gar an eiē kakon tou agathou to homoiōma?* De oratione dominica: I PG 44: 1132 B.

41. See Jaeger. *Two Rediscovered works* pp 26 ff.

42. *Ho kat' aretēn bios* (the life according to virtue) is a platonic term (see eg. Theaeteus 176 B and Republic: 613 B) which Gregory regarded as central. It occurs in the sub-title of Gregory's *Life of Moses, de Virginitate*, and is central in *De Perfectione.*

> that it does (through us). Neither is it likely that the good (nature) would activate something beside its own nature"[43]

It is for this reason that Gregory regards the true action of the human nature to be virtue. Human nature is acting in conformity with its own being only when it practices virtue. When it practices vice or evil, it is acting in accordance with this "non-existent" alien power, evil or sin.

All good action has its original cause in God's goodness and is therefore the result of grace. But only when man acts in freedom to perform the good, does he become truly man, acting in conformity with the good, in freedom. If he were to do good in a merely passive response to a sovereign grace, then man can be good like a flower is good or a river is good. Only freely chosen good actions, or in our language, the moral good, is truly human good. And that cannot be a passive response to a sovereign grace.

Virtue, or acting that which is good, is acting in accordance with human nature, the essential character of which is neither to sin, nor passively to be moulded by a sovereign grace, but rather to be good by free choice. In this sense, virtue is the true nature of man, and the basis of man's participation in God's goodness.

Virtue is thus the true character of the new man, who puts on Christ, and this means to put on love, holiness and righteousness.[44] It is the new garment, the garment of immortality that makes us human again, free agents of the good, akin to God.

It is this same virtue, or growth in the practice of the good, that becomes also the knowledge of God. In a classical passage commenting on the Beatitudes, Gregory makes his point clear:

> "Now the divine nature, as it is in itself, according to its essence, transcends every act of comprehensive knowledge,

---

43. In Inser: Psalm. I:VIII. GNO V:53/17-23 PG 44:468 B-6. The meaning of the last sentence is somewhat obscure: *Oude gar eikos ēn ton agathon tēi phusei allo ti par' ho pephuken energēsat*. The meaning seems to be that the nature of man, which is created as good, cannot do evil, and that evil comes from outside the nature, though evil itself has no nature of its own, not being created.

44. *In Canticum. Cant.* Hom XI. PG 44:1005 A-1007 A Hom XIII PG 44:1044 B GNO VI: 328/1-333/5; 378/12-21.

and it cannot be approached or attained by our speculation. Men have never discovered a faculty to comprehend the incomprehensible, nor have we ever been able to devise an intellectual technique for grasping the inconceivable. For this reason the great Apostle calls God's ways *unsearchable* (Rom. 11:33), teaching us by this that the way that leads to the knowledge of the divine nature is inaccessible to our reason; and hence none of those who have lived before us has given us the slightest hint of comprehension suggesting that we might know that which in itself is above all knowledge.

"Such then is He Whose essence is above every nature, invisible, incomprehensible. Yet He can be seen and apprehended in another way, and the ways of this apprehension are numerous. For we can see Him, Who has *made all things in wisdom* (Ps. 103:24), by the process of inference through the wisdom that is reflected in the universe. It is just as in human works of art, where the mind can in a sense see the author of the ordered structure that is before it inasmuch as he has left his artistry in his work. But notice that what we see here is not the substance of the craftsman, but merely the artistic skill that he has impressed in his work. So too, when we consider the order of creation, we form an image not of the substance but of the wisdom of Him who has done all things wisely...For being by nature invisible, He becomes visible only in his operations, and only when He is contemplated in the things that are external to Him.

"But the meaning of the Beatitude (Matth. 5:8) does not merely indicate that we can infer the nature of the cause from its operations; for in that case even the wise of this world might gain a knowledge of transcendent wisdom and power through the harmonic structure of the universe. Rather, I think this magnificent Beatitude envisages another counsel for those who are able to receive and grasp what they desire...

"The Lord does not say that it is blessed to know something about God, but rather to possess God in oneself: *Blessed are the clean of heart, for they shall see God* (Matth. 5:8). By this I do not think He means that the man who purifies the eye of his soul will enjoy an immediate vision of

God; rather I think this marvelous saying teaches us the same lesson that the Word expressed more clearly to others when He said: *The Kingdom of God is within you* (Luke 17:21). And this teaches us that the man who purifies his heart of every creature and of every passionate impulse will see the image of divine nature in his own beauty. So too in this short sentence the Word, I think, is giving us the following advice: All you mortals who have within yourselves a desire to behold the supreme Good, when you are told that the majesty of God is exalted above the heavens, that the divine glory is inexpressible, its beauty indescribable, its nature inaccessible, do not despair at never being able to behold what you desire. For you do have within your grasp the degree of the knowledge of God which you can attain. For, when God made you, He at once endowed your nature with this perfection: upon the structure of your nature He imprinted an imitation of the perfections of His own nature, just as one would impress upon the outline of an emblem. But the wickedness that has been poured all over this divine engraving has made your perfection useless and hidden it with a vicious coating. You must then wash away, by a life of virtue, the dirt that has come to cling to your heart like plaster, and then your divine beauty will once again shine forth.

"Once he has scraped off the rust-like dirt that has accumulated on his form because of evil degeneration, then will he become good once more and shine forth in the likeness of his archetype. For surely what resembles the Good is in itself Good. Thus if such a man will look at himself he will see within himself the object of his desire, and thus he will become *blessed*, for in gazing upon his own purity he will see the archetype within the image.

"It is just like men who look at the sun in a mirror. Even though they do not look up directly at the heavens, they do see the sun in the mirror's reflection just as much as those who look directly at the sun. So it is, says our Lord, with you. Even though you are not strong enough to see the light itself, yet you will find within yourselves what you are seeking if you would but return to the grace of that image which was established within you from the beginning. For the Godhead

> is all purity (*katharotēs*), freedom from passion (*apatheia*) and separation from all evil (*kakou pantos allotriōsis*). If these qualities are in you then God is surely within you, when your mind is untainted by any evil, free of passion, purified of all stain, then will you be *blessed* because your eye is clear. Then because you have been purified you will perceive things that are invisible to the unpurified. The dark cloud of matter will be removed from the eye of your soul, and then you will see clearly that blessed vision within the pure brilliance of your own heart. And what is this vision? It is purity, holiness, simplicity, and other such brilliant reflections of the nature of God, for it is in these that God is seen"[45].

Here is the true knowledge of God, the blessed vision, or beatific vision. It is not at all a "mystical" experience, but something which happens when man becomes truely human, untainted by evil, in full control of oneself, grown in goodness and holiness. It is certainly impressive to see how Gregory unites the intellectual or cognitive, with the so-called mystical and the so-called ethical. Knowledge itself becomes an act of love, inseparably linked with holiness and true inner liberty. As Gregory says clearly in *On the Soul and the Resurrection* "Knowledge becomes Love"[46]. *Apatheia* is an essential element of that love, but not in the sense of the absence of passion; rather, love is that strong self-directed act of the will which is dispassionate only in the sense that the human will is not rendered passive or overwhelmed by some force over which it has no control. Passion for Gregory means passivity i.e. to be the *object* of an action on feeling, rather than the subject who retains free control of the movements in the action. Sovereign grace, as conceived by Augustine, would make man too passive, and therefore less than human; so Gregory rejects the idea of a sovereign, compelling, dictatorial grace. Greater than a compelling grace is the grace that allows man to be a free source of the good. The gift of freedom is a greater grace than the most sovereign grace. To love the good freely and not by compulsion—that is the greatest gift of God, and this gift is given to Man. It

---

45. *De Beatitudinibus, oratio VI.* PG 44:1268B-1272C Eng. Tr. from Musurillo, *From Glory to Glory* pp 98-102 (slightly altered).

46. *de anima et resurrectione* PG 46:96C *hē de gnōsis agapē ginetai.*

is precisely because Man is a Middle Reality between God and the Creation and is consequently drawn in both directions at once, that there arises the opportunity to exercise his freedom.

> "Since our nature is something of a double character (*diple tis estin*), commingled of the intellectual and the sensual, logically our life is also double, in tension between these opposing elements, corporeal in the sensual part, incorporeal and intellectual in the other. So also each of these parts of us has an idea of the good which is not one and the same. The intelligible part looks towards the intelligible good, whereas the corporeal and sensible part desires the sensible good"[47].

It is in the context of this struggle between two opposing propensities that man is asked to discern and choose that which is truly good. The one thing that is truly Good is God, and when our passionate and self-directed will finally fixes itself upon God as the one good to be chosen above all others, than we begin to see God and also become God.

Here love of God becomes the true expression of human freedom. In this act of love is also the true act of knowledge, for in the love of God one gets to know the reality that lies behind all reality, the One that is alone good and true and reliable.

> "For different are the truly good things (from what appears good to the senses)—the things which are good in themselves and make good those who participate in them. It is of course possible for something which participates by nature in another to be transformed in the direction of that in which it participates. For example the mouth of someone receiving a sweet-smelling spice in the mouth becomes itself sweet-smelling, and again someone else who chews garlic or something even more bad-smelling develops a foul-smelling mouth. So since the foul smell belongs to the filth of sin, the opposite of it, namely the virtue of Christ (*hē aretē Christou*) is fragrance, i.e. the loving relationship effects a natural commingling with the beloved (*hē de agapētikē schesis tēn pros to*

47. *In Ecclesiasten*: Oratio VIII. GNO V:419/9-16.

> *agapōmenon anakrasin phusikōs katērgazetai*). Whatever therefore we choose through our love, that we also become—either the fragrance of Christ, or the foul odour (of sin). For he who loves the good shall himself become good, the goodness that is contained in him transforming the container itself into itself"[48].

For Gregory of Nyssa there was no conflict between grace and human effort. It is grace which makes human effort possible. The grace of Christ in the Incarnation now acts as a special means of grace, drawing human beings to the love of the good. The sacraments also confer grace. But in no case can grace become compulsive, for then it can no longer effect the truly morally good, for "virtue compelled is not virtue"[49].

Without Christ, human nature, the "sheep gone astray" could not have by itself returned to the Shepherd[50]. So without that grace humanity could not have produced any virtue. Grace in creation, grace in redemption—both these are accepted as fundamental realities by Gregory. Through Christ, i.e. "in the new man created in the image of God, in whom the fullness of the Godhead dwelt bodily", who through his own purity rendered the human nature again akin to God, "he has drawn us all together into the same grace—the whole human nature that now shares in his body and has been made akin to Him"[51].

The difference between Augustine and Gregory at this point is rather subtle but significant. Augustine does state clearly that all nature is good because created[52]. He also states that the works of a sinful person could not redeem himself[53]. With these Gregory would readily agree. But Gregory would say that man has the capacity to repent and to turn towards God, though this turning too could be interpreted as grace. Where Gregory genuinely differs

---

48. *In Ecclesiasten:* Or VIII. Gno V:422/12-423/4 PG 44:737C-740A.

49. *De opificio hominis* XVI. PG 44:184B NPNFV :405B *Adespoton gar ti chrēma hē aretē kai ekousion, to de katēnagkasmenon kai bebiasmenon aretē einai ou dunatai.*

50. *Contra Eunomium* Bk. III. Tom X GNO II 293 ff PG 45:889 B NPNF V:241 B.

51. C.E. *Loc. cit.*

52. *Enchiridion* IV:12; 13.

53. *Ench*: IV:30.

from Augustine is in insisting that human acts of freedom, with the co-operation of the Spirit of God, can decisively shape one's own being, and that while our works may not be capable of redeeming us from our lostness, still these works of ours have a decisive force in shaping the structure of our own being. Gregory seems to put the emphasis on the freedom of the redeemed human being, while Augustine places the emphasis on the unfreedom of the unredeemed.[54] Augustine fears to dwell upon the positive aspects of the freedom of the man redeemed in Christ. Gregory glories in the freedom of the redeemed humanity to practice virtue. Augustine used strong language when he says that the *gratia Dei per Iesum Christum Dominum nostrum* is absolutely essential for our liberation from evil[55]. Gregory would agree. When Augustine goes on to say about the grace of God in Jesus Christ, that without it human beings can "do no good whatsoever, either in thought, or in will and love, or in action" (*Sine qua nullum prorsus sive cogitando sive volendo et amando, sive agendo faciunt bonum*), he gives rise to the possibility of the mis-interpretation that only Christians can do good. The other option is to say that the "Grace of our Lord Jesus Christ" is freely available to non-Christians, since they do in practice do good. Gregory of Nyssa would not make the mistake of saying that without the grace of God as mediated in Jesus Christ no man is capable of doing any good. He clearly cites Moses and Elijah as people who used their freedom in the right way, rather than as mere recipients of a special grace not received by others.

The difference though rather subtle, is exceedingly important today for us in dealing with the realities of secular societies and in dialogue with other religions.

The argument of the Pelagian monks at Hadrumetum (North Africa) is summarized for us by Augustine in para 6 of *de correptione et gratia:*

> "They say, if God gives the grace to do good, then clearly my not doing good is due to God's not giving grace. So you should not admonish me, but pray that effective grace be given to me.

54. see e.g. *de correptione et gratia* ch. 1.
55. *Ibid* Ch II:3

> There is no use admonishing me, because, if the gift is not there, what can I do?"[56]

Augustine in his reply evades the question by saying that the one who is unwilling to be admonished, should for the reason of his unwillingness be admonished. Augustine goes on to make out a case why such a person should be admonished; but he does not say what meaning admonition could have if the will of God is sovereign, and if the grace of God is capable of producing its own effects without any admonition.

Gregory takes a different line. Augustine prefers the image of Potter and Clay. This is scriptural. But if pressed by itself, it does violence to the freedom of man. Gregory cannot take the line that man is mere clay. He is clay on which the stamp of the freedom of God has been impressed. The freedom has been given to humanity as such by the grace of God in creation. Gregory would be reluctant to limit grace to the dispensation of the incarnation. All humanity, by virtue of their created nature, share in the capacity to do good. The fact that most people fail to do so does not prove the absence of freedom. On the contrary, the fact that a few do practice good, is evidence enough for the presence of freedom in all human beings.

For Gregory, grace is primarily manifested in the great acts of God and in the sacraments. He speaks about the "grace from above"[57], the "grace of the resurrection"[58], the "grace of incorruptibility"[59], the "grace of the Spirit"[60], the "grace of our Lord Jesus Christ"[61], the "grace in ordination"[62], the "grace in Baptism"[63]. Especially in the homily *On the Day of Lights,* one finds frequent reference to 'the grace of washing', the royal grace (*charis basilikē*)

56. Eng. Tr. John Courtney Murray, series, *Fathers of the Church, St. Augustine, Admonition and Grace.* Writings of St. Augustine Vol. IV N.Y. 1947. p. 250.

57. *tēs anōthen charitōs. In Diem Luminem* GNO IX:227/7 PG 46:584 B.

58. *hē tēs anastaseōs charis.* Epist V:5 GNO VIII:II:32/27—33/1 see also 94/7-8.

59. *hē charis tēs aphtharsias* Epist V:6.

60.. *tēn charin tou pneumatos*—Epist XII:3. Ibid 44/1-2 see also GNO IX:233/2 PG 46:592 B.

61. Ep. XVII:I Ibid 51/6-7.

62. Ep. XVII:28 Ibid 57/17-20.

63. See passim *In Diem Luminem* GNO IX:223/2, 230/9 etc.

which renews our old man and recreates us, which makes us participants in the grace of the death and resurrection of Christ. For Gregory grace par excellence connotes not the help God gives for performing individual good deeds, but rather God's great intervention in the Incarnation of Jesus Christ to redeem man from sin and death. This is particularly the grace of the Spirit, but Gregory insists that it is equally the grace of the Trinity. Grace is a mighty act of God, not some petty help given to each man from time to time. It is manifested in the conversion of Paul, who before grace was a persecutor, but after grace becomes an Apostle.[64] The message of the gospel is the annunciation that the grace of God has appeared in Jesus Christ.

It is this element which Pelagius had inadequately grasped. It is only because the grace of God has appeared in Jesus Christ and renewed the whole of human nature in one single act of redemption, that human beings now have the possibility of practising "virtue" and living in the good.

Gregory emphasizes the grace of God in creation, the grace of God in the Redemption through Jesus Christ, and the grace of God in the mysteries of Baptism, Eucharist, Chrismation, Priesthood etc. He does not place as much emphasis as Augustine does on God's help for every single good act of man. On the contrary the good acts of man are the proper response to the grace of God, who has created man to be a *free agent of the good,* and when man lost that capacity, restored it to him in Jesus Christ. It is the grace of creation, grace of redemption and the sacramental grace that make it possible for man to do good at all, but the good has to be a free act of man and not something compelled by a sovereign grace.

Pelagius was probably not subtle enough to understand this aspect of Gregory's thought, and it would be more legitimate to call Pelagius a mis-interpreter of Gregory's thought than to call Gregory a "semi-Pelagian". There is a great gulf fixed between Gregory and Pelagius. Gregory's thought is fundamentally Christian, inseparable from the Christian faith. Pelagius' thought seems fairly independent of Christian teaching.

Gregory's doctrine of *synergia* is an essential element in his understanding of Man as Image of God, which means the capacity

---

64. *In Diem Luminem.* PG 46:596 D. *Dioktēs ho Paulos. alla meta tēn charin Apostolos.*

to do good and be good. Gregory might have insisted as Pelagius seems to have done that it is only the *posse*, the possibility for doing good, that God provides. Man's *velle*, willing, is an essential element in the act of virtue. God does not do historical acts of virtue except through the human agency. Therefore the human agent is an essential element in history. What man does in his freedom, redounds to the glory of God. If man creates a robot who does some things which man can do, man's glory in that act is less than what his glory would have been if he could create a living being who can do things which man cannot do by himself. God's glory is greater when man, His creation, is able to do great things. There is no conflict between God's glory and man's. The more marvellous the creation becomes, the more glory its Creator gets. It is in this context that Gregory may have held the view, if the *Instituto Christiano* is authentically Gregorian, that divine grace is proportional to human effort. But that way of putting it is perhaps too strong. What Gregory does say (if it is Gregory who says it) in the *Instituto Christiano* is this :

> "The growth of maturity of the body does not depend upon us . . . But the measure of the soul, in the renewal of its birth and beauty, depends on our judgement; i.e. the good we have through the effort taken by us with the help of the grace of the Spirit. As far as we stretch the striving and effort on behalf of the true religion, so much the stature of our soul grows through our striving and effort".[65]

One can see clearly here that what takes the place of grace (as understood by the west) is in Eastern thinking, the doctrine of the Holy Spirit. Authentic Christian thinking does not think of grace as a special entity with a hypostasis of its own. Grace is the action of the Holy Spirit, and though this is "unmerited", it is not a thing in itself.

Jaeger gives us a full account of all the passages in *de Instituto Christiano* which relate to the concept of *synergia*.[66] He has however failed to relate fully the idea of divine help to the personality

---

65. *de Instituto Christiano* P. 46:7ff. See Jaeger, *Two Rediscovered Works* p. 93. Callahan, *Ascetical works*. p. 131.

66. pp 93-96.

of the Holy Spirit. This grace is often referred to as the grace of the Spirit (*hē tou Pneumatos charis*)[67], the power of the Spirit (*hē dunamis tou Pneumatos*)[68] the help (supply) of the Spirit *hē epichorēgia tou Pneumatos*)[69] etc. It is clear that for Gregory all human acts of good are God's acts not only in the sense that God has given man the capacity to do good, but also because the Spirit, i.e. God Himself, is in person present to help him to do good. It is the Person of the Spirit rather than an abstract entity called grace that helps man do good. But what God gives is only possibility and help. The agent is man. Otherwise it would not be a free act. In the western doctrine of grace, God becomes the agent, and our own wills only cooperate. Here in this catholic teaching as represented by the thought of this "universal father", God is *before* the act of good as the One who gave man the possibility of doing good, and *in* the act of doing good as helper, but never assumes the role of agent, for that would be to deny the freedom of man. Man can perform acts of good which seem to be beyond the normal capacity of human nature.[70] But man is still the agent.

The gift of eternal life is indeed a gift of the Spirit, Gregory affirms; it is the Spirit of God who gives this gift, but we acquire the worthiness or the capacity to have and to enjoy this grace, through faith which is manifested in the tireless capacity to strive for the good.[71]

This special power of the Spirit to generate and enhance the capacity for good actions is what the ancient tradition of the Church regards as grace. It is an act of God, but God's agency does not overwhelm the agency of man. Only a proper doctrine of the Holy Spirit can hold together an authentic doctrine of grace and a full understanding of man as a free being.

The receiving of the Spirit is not an arbitrary and capricious matter, however. The Holy Spirit acts more effectively in those who have made themselves holy—negatively by separating from sin and

67. p 46/6, 10, 26, 55/14, 57/14, 61/21, 77/11 etc.

68. p 45/16, 54/14.

69. 58/14-15.

70. Here we find one of the few instances in which Gregory uses the expression, almost meaning 'supernatural'—*ta huper tēn phusin autōn agatha*—' "good deeds beyond their nature".

71. 46/26-28.

positively by practising acts of virtue[72]. The Holy Sprit's special capacity is to do good through the will and agency of man without destroying man's freedom. On the other hand,

> "do not acquiesce in His gifts, thinking that because of the wealth and ungrudging grace of the Spirit nothing else is needed for perfection. When these riches come to you be modest in thought, ever submissive and thinking of love as the foundation of the treasure of grace for the soul, struggle against all passion until you come to the height of the goal of true religion to which he leads his disciples through prayer and teaching, showing to those who love the Lord the change for the better and the grace which results from love."[73]

Grace as help for good deeds, is the consequencc of the indwelling of the Holy Spirit, and that indwelling does not take place until one has become a worthy abode for the Holy Spirıt through seperation from evil and the practice of virtue. The Holy Spirit comes and dwells in those who fight the other loves that distract man from the love of the good, and who take up their cross with joy and hope. Gregory will grant no cheap grace.

But then neither does Gregory want anyone to boast or even be conscious of his own virtues:

> "The lover of grace from on high endures every labour for the sake of those things which attract the Spirit and, having obtained a share in the grace from that source, he produces fruit and enjoys the harvest which the grace of the Spirit cultivates in his own humility and active zeal. It is necessary to endure the toils of prayer and fasting and the other works with much pleasure and love and hope, and to believe that activities are the flowers of labour and the fruits of the Spirit. If anyone imputes those things to himself and gives himself entire credit for them, in place of the undefiled fruits there grows up in such a person false pretension and pride, and these passions, like some blight growing in the souls of those easily satisfied, destroy and nullify the labours"[74]

72. 61/19-22.

73. *De Instituto Christiano* Eng. Tr. In V.W. Callahan, *Saint Gregory of Nyssa, Ascetical works* p. 141.

74. *de Instituto Christiano*. Callahan p. 155,

The Spirit, giver of grace, does not bestow his gifts in such a way as to abrogate or overwhelm the freedom of man. But the Spirit's presence in Man and His willingness to work through and in man together constitute one of the great mysteries of our present-day existence, particularly since the Incarnation and Pentecost. The one great consequence of God's becoming man is the fact that the Spirit is now present in the community of Faith and Holiness. The Spirit dwells in Man, inspires both his knowledge and his action, and transforms his very being—but all this neither as a servant, nor as an arbitrary dictator, but as One who works in a mysterious way as God's Presence in the whole of creation, but particularly in Man.

Man's freedom, however, has to be used even for providing the set-up in which the Spirit can work. When Man separates himself from evil, attaches himself to God, lives a life of holiness and service. and dedicates himself unremittingly to prayer and virtue, then the Spirit abides in Man and Man becomes the Presence of God in creation—which is what it means to be the Image of God. As Gregory says:

> "The husbandmen of Christ and truth who through faith and the toils of virtue, have received goods from the grace of the Spirit beyond their nature, harvest with unspeakable pleasure, and without effort they attain a guileless and unshakeable love, unmoveable faith, unfailing peace, true goodness, and the rest of the things through which the soul becomes stronger than itself and more powerful than the evil of the enemy, and furnishes itself as a pure dwelling place for the Holy and adorable Spirit. From the Spirit, it receives the eternal peace of Christ and, through it, unites with and cleaves to the Lord."[75]

The Spirit thus unites the human person to Christ, but only when Man has separated from evil and cleansed himself, through baptism and the Eucharist on the one hand, and through the intense self-discipline of the love of God on the other. But the will of man is a necessary precondition for the perfection of man. The grace of the Spirit, coming to us in the preaching of the Word, in the water of baptism and in the heavenly food of the Eucharist

75. *de Instituto Christiano* Callahan p. 156.

has to be perfected through a life of virtue. For this life of virtue, as well as for the bearing of the cross which is the inescapable outcome of the life of virtue, we need the grace of the Spirit, but this grace is not a hypostatic entity, but the presence of the Spirit in our bodies and souls energising us and strengthening us beyond our normal capacity to greater degrees of good.

It is a great misunderstanding of the authentic tradition to think of *synergia* as *our* cooperation with the grace of God. *Synergia* refers on the contrary, to the help and assistance given to the Christian by the indwelling Spirit in order to perfect the life of freedom and virtue. This was not what Pelagius had talked about. Pelagius had grasped the idea that our wills are important in the life of virtue, but had not understood the whole Christian context in which the Bishop of Jerusalem had taught.

In Augustinian (perhaps not Augustine's own) thinking, grace is necessary, because our nature is evil. This Gregory definitely repudiates:

> "If man is born of evil, how could any deliberate impulse towards the good life be present in a man, since his nature, as they say, has its beginning in evil?............Therefore on account of this, it has clearly shown that evil is not older than life, and that nature does not have its first beginnings in it, but that the wisdom of God, which governs everything, is the beginning of our life."[76]

For Gregory the redemption in Christ restores unto man the possibility of being and doing good. As he separates from evil and cleaves to God in Christ by the Spirit, the Spirit indwells him and empowers him. It is this empowering that enables him to be transformed into the image of God. But the empowering is not a thing called grace, but the presence of God and the power of God. But the presence and the empowering of the Spirit take place without annulling the agency of the human will. It is still man's act, but with God's power. Only in the absence of a proper doctrine of the Holy Spirit is there need to develop an elaborate doctrine of total depravity and sovereign grace, both of which mark departures from the authentic tradition of the Christian Church.

---

76. *de Anima et Resurrectione: Callahan, Ascetical Works, pp 251-252.*

CONCLUDING POSTSCRIPT

1. God-World Relationship

In any discussion of the God-Man-World relationship two fundamental points have always to be kept in mind. Those two points were well understood by Gregory of Nyssa, and they have also to be applicable to Gregory's thought.

1. In the first place, the concept of *diastēma* implies that what lies beyond the *diastēma* is incomprehensible in terms of the *diastēma*. Our conceptual knowledge is limited to the boundaries of the *diastēma* or time-space creation. If this is so, our conceptual language itself is diastematic. We cannot use this language to describe what lies beyond the *diastēma*. That is clear in Gregory of Nyssa. But can we then use that language to describe the *relationship* between what is in the Creation and what is beyond?

There are obviously three difficulties—First, one of the two terms of the relation lies beyond the limit of our comprehension, and even the other term, that is the diastematic creation itself, is not *fully* comprehensible in as much as we are unable to stand outside that creation to gain perspective of the creation as a whole. Thus one of the two terms of the relationship we cannot comprehend at all, and the other we comprehend only partially, i.e. from a point of view inside the reality that is to be comprehended and not from the usual distance necessary for the subject-object relationship. Thirdly, even if we had a vantage point from which we could see the diastematic creation as a whole, since the diastematic creation is not a distinct entity in itself but a contingent reality dependent for its very existence on the Creator, it cannot be comprehended as a clear and distinct object in itself. Any understanding we can thus have of the relation between Creation and Creator has to be necessarily indistinct.

In view of these difficulties we can already say that any statement we make about the God-World relationship can be only of a tentative, hypothetical, almost mythological nature, and not an accurate account in analytical or descriptive terms.

2. Secondly since our concepts, language and logic are adapted to description of relationships within the diastematic or time-space creation, it is practically futile to use these for any accurate description of the relationship of the Creation to that which is beyond the *diastēma*. We should not deceive ourselves into thinking that any account, including that of Gregory of Nyssa, gives us a full and accurate description of that relationship which will always elude our conceptual or linguistic grasp.

3. There is the further problem that the very notion of *diastēma* used in this way to denote a gap, screen or separation between the Creator and the Creation, is itself a spatiotemporal concept and therefore essentially unclear when used for the relation between God and His Creation. This applies also to the concept of *metousia* which Gregory uses to refer to the same relationship. Both *diastēma* and *metousia* are concepts which have to be understood in an evocative manner, and not as descriptions of that relationship.

What then is the purpose of such concepts? They do serve the function of controlling notions by which to check error. They contain sufficient reflection of truth as to be used for verifying the reliability of certain other statements. Let us take a few of these examples.

(a) *Where is God?* Any statement that implies that God is located at a particular point in space can immediately be questioned on the basis of our understanding of time-space and *diastēma.* The being of God cannot be in a particular point in heaven or earth, and all statements which have such time-space reference would then have to be understood in a suitably qualified manner—the throne of God, sitting at the right hand of God, the angelic choirs around the throne of God,—all these will have to be conceived in non-spatial terms, if such conception were possible.

Even more problematic is the concept of number. When we say God is three-in-one, are we not making a time-space conceptualization and applying diastematic logic to the Creator's being? The Trinitarian conception of God certainly goes beyond the *energeia* of God and is normally meant as a confession of the nature or *ousia* itself of God. We say Three Persons in one *Ousia* about God who is beyond the *diastēma.* Are we not applying a diastematically limited conceptual language to the very being of God in this case by introducing concepts like number into the

God-head? Can we count one, two, three in relation to the God-head, for two can come only after we have reached the boundary between one and two, but if one and two have no boundary, how can we reach two at all, starting from one?.[1]

If we say there are three persons in the Trinity, would they not have to be finite beings in order that we can count the number three? If this is so, then is not the Trinitarian doctrine itself imperilled by the notion of *diastema?*

(b) *Secular World?* If we accept the scheme that all created being is contingent, and is sustained in existence only by the *energeia* of the Creator, then are we justified in resorting to a secular interpretation of the cosmos, as if the cosmos were self-contained and could be explained in terms of itself? Can there be a non-metaphysical or non-ontological understanding of this world for a Christian who understands the doctrine of creation not merely in the sense that at the beginning of creation there was an act of God, but also in the sense that the existence of the creation now is the continuing manifestation of the *energeia* of God? Does not the biblical claim that the things that are visible (open to our senses) have come to be from the unmanifest (*me ek phainomenon*—Heb. 11:3) make much better sense than the secular claim that this world now open to our senses is the only world that there exists, and that the Gospel has to be understood only in terms of this present world of our time and our space? If we sincerely reject the secular claim, then would we not have to take a second look at our "political theologies" and our socio-economic understanding of liberation and salvation?

(c) *The Role of Negative Theology.* We can make affirmative statements about God, only if we keep negative theology always as a corrective and counter-balance to any positive affirmations we may make about that which lies beyond the *diastema.* All theological statements, in so far as they contain any reference to

1. Gregory began dealing with this question of number in a creative way but did not pursue it to any clear logical conclusion. If number is defined as a 'synthesis of monads' (*In Hexaem.* PG 44:85 C—*Ouden gar heteron estin arithmos, ei mē monadōn sunthesis*, and if a monad is by definition bounded (*to diōrismenē tini perigraphēi-ibid*), then how can the number three be applied to the Trinity at all? If number is the sign of the quantity of known realities, (*Contra Eunom* I. PG 45:312 B) how then can the Trinity be numbered at all?

that which lies beyond the *diastema*, (and clearly one cannot discuss Trinity, Creation, Incarnation or Eschatology without such reference) have to be suitably qualified by a negative statement to the effect that this *kataphatic* or affirmative theological statement is only a necessary over-simplification in conceptual terms of that which is conceptually unstateable. Negative theology or apophatic theology cannot then be regarded as merely one of the many ways of doing theology as is implied in the term *via negativa*, but a necessary element in all positive theological affirmations. Gregory of Nyssa says this in his *Answer to the Second Book of Eunomius*.[2] Things that we do not know, we cannot coin words by which to signify them. The things that come within our comprehension are of four kinds:

(a) existents conceptualized as existing in diastematic extension.
(b) existents conceptualized as having their own spatial capacity within which other things are contained.
(c) existents coming within our vision as temporally conditioned, with a beginning and an end, with non-being beyond in both directions.
and (d) phenomena grasped in terms of changing qualities—disintegration, suffering, change and alteration in different combinations.

All we can do, says Gregory, is to say whenever we make a theological statement, that these four elements do not apply to the Transcendent Being. We create negative terms like pre-temporal (*proaiōnion*), unbegun (*anarchon*), unending (*ateleutētón*), incorporeal (*asōmaton*), incorruptible (*aphtharton*), passionless (*apathes*), unchanging (*atrepton*), unalterable (*analloiōton*). Not because we have direct conceptual experience of such realities. Our experience is only the opposite of these. But yet we dare to posit a minus symbol, as in mathematics we dare to speak of a 'minus ×' as if it were an existent reality. So most of the things that we can say about God are in these negative terms; but they do not give us any positive conceptual hold upon the nature of God. As Gregory says, only the operations of God can be spoken of in affirmative terms, but the being of God must be conceived only

2. PG 45:909 ff. Migne's *Liber XII.* GNO I:226-409, Jaeger's *Liber* See especially PG 45:1101 D ff where Gregory starts on a long epistemological discourse.

in negative terms. We say Creator, Almighty, Ruler, the Supreme Good, but in none of these affirmations we should think that we have a positive understanding of the Transcendent Being, for it is not in the way *we* create, rule, or do good that *He* creates, rules, and is good.

That is why even when making operational statements about God, it is not enough to use positive analogy; it is necessary to make the negative analogy—not as we create, not as we rule, not as we are good.

This is an important consequence for contemporary theological reflection, that we have to be aware of the fact that all our statements about God's operations would also have to be qualified by this negative theology. When we say that God acts in history, or that God's will is revealed in the Scriptures, or that God is the ruler and judge, these statements should be made only with that negative qualification.

2. **Man-World Relationship**

What insights does our analysis of Gregory's thought yield about the Man-World relationship?

Here Gregory affirms both continuity and discontinuity. The continuity can be briefly analysed as follows:

(a) *Both Man and World are created,* and are parts of the diastematic or created order of reality. Both have come to be from that which was not, and both continue in existence only contingent upon the will of the Creator, by His *energeia.* Both are extended in time and space, and share in the situation of having an *arche* (beginning) and *telos* (goal) and a way to traverse in between. Both are corruptible, changeable, impermanent.

(b) *There is a basic continuity between matter, organic life and human life.* It is the same creative *energeia* of God which has brought matter into existence, and through that same creative impulse brought into being successive stages of existence—the vegetable kingdom, the animal kingdom and the human realm. This Gregory affirms very strongly in *On the Making of Man.*

But he also affirms that humanity is more than merely continuous with the rest of creation. Mankind or human nature holds within itself all the lower kinds of life. This is not the same as saying that man is a microcosm of the world. The pagan idea of man as microcosm was based on the fact that he is constituted of the same four elements as the earth is made of. This idea Gregory rightly rejects on the ground that gnats and lizards are also made of the same four elements and therefore they would also be a microcosm. Gregory's position is that humanity contains within itself all the lower forms of soul—the vegetative, the animal or sensitive and the rational or intellectual. Man is in fact superior to the angels in this sense that he is commingled of the material and the intellectual. "Nature advanced in an orderly assent to perfection" and the finally evolved form, humanity, comprises in itself the lower forms through which it has ascended.

Thus the basic continuity between man and the world is not merely in that both are made up of the same four elements, but rather that man is capable of being the highest evolved recapitulative centre of the universe.

(c) *Discontinuity between Man and World.* This basic element of continuity is to be held together with the element of discontinuity between man and world. The discontinuity lies precisely in the fact that Man is the image of God. The meaning of his being is not to be explained in terms of an evolutionary anthropology alone, though that is an integral part of his being. Its meaning is not to be derived primarily from the fact that human nature recapitulates in itself all lower forms of life as well as inorganic matter. The true character of human nature is revealed only in the insight that Man is made in the image of God.

(d) *Implications.* The relationship between God and Man we will reserve to the next section. But here we need to dwell a little longer on the implications of the Man-World relationship.

Gregory would be the first to insist, along with the whole patristic tradition, that man is born to be king of the universe, ruling it as lord. Gregory says so at the very outset of *On the Making of Man* (II:I to V:2). But Gregory's teaching about the lordship of man over creation does not make the rest of creation merely an object to be analysed and manipulated by man according to his own wish. That is an understanding of sovereignty developed in certain countries of the world where the sovereign Lord is a despot whose every command the rest of the world is required to obey without question.

Gregory would not allow humanity to 'enjoy the creation' without in the same act 'enjoying the Creator'. Humanity's essential nature is double—to enjoy the created world and the uncreated energies of the Creator at the same time. A secular understanding of humanity as master of the creation thus becomes essentially one-sided, and is a betrayal of the true nature of man. Humanity can truly enjoy the sensible creation only in so far as it cultivates the capacity to enjoy the Creator's presence and love.

Science and technology as we now have them, have grown up in a world where secularization as a process has cut off man and world from awareness of their transcendent source and ground. Therefore the patterns of lordship developed by science-technology have taken shape in a distorted human context. The inseparable trio is Creator-Man-Creation, man being the participant mediator between Creator and Creation. And in so far as science developed in a secular context, only two elements were taken into account—namely Man and the rest of creation.

Man who exercises lordship over creation without reference to his communion with God and to his contingent existence dependent upon God as Creator, is distorted man, Gregory would say. And therefore his lordship over creation would also be distorted and dehumanizing.

Man is not master of the world on his own. He can become truly master of the creation only by being related to the Creator as image or manifest presence.

This means that we will need to develop a 'science' and 'technology' that will keep our relationship with the other pole of our existence—with our Creator and our archetype, God. This is what Faith and Worship, with life in the community of the Spirit truly means. The life in the Community of the Spirit or the Church is the parallel, nay even the foundation, for that other reality which we partly inherit and partly create—namely our civilization. What we have inherited and are fostering is a civilization which has definite roots in the Community of the Spirit. Neither Marxism nor liberal democracy can completely disavow all debts and obligation to the Community of the Spirit, often wrongly called the Judaeo-Christian heritage.

But we are in a situation where there is a growing repudiation of that heritage coupled with deep-seated mis-understanding of the meaning of life in the Community of the Spirit.

In other words as the civilizational pole of our existence has been further developed by the phenomenal growth of science and technology, we have, instead of deepening our rootage in the spiritual pole of our existence by more perceptive participation in the Community of the Spirit, tended to distort, neglect, reject, betray, falsify and misinterpret that community.

It is here that we desperately need a new science and technology for deepening our participation in the spiritual pole of our mediatorial existence as humanity between God and His Creation. This new science and this new technology should clearly be a new way of experiencing our relationship to God in Faith, Love and Hope, through the life of worship, discipline and service.

In prospecting for this new science of Faith, Hope and Love, and in developing the new technology of living out that life in worship, discipline and service, we may find the whole neglected patristic heritage of the Church to have a necessary first priority of claim on our attention. It is in that context that we need to take a deeper look at Gregory of Nyssa and his way of thinking and experiencing reality.

3. The God-Man Relationship

In the relationship of God and Man lies the most profound mystery of our existence. Words are unlikely to penetrate the depth of that mystery. Yet Gregory of Nyssa has opened some new perspectives from which to peep at the outskirts of that mysterious relationship.

The concept of Image has been duly studied by many. In the case of Gregory, Fr. Leys' study of 1951 (*L'Image de Dieu chez Saint Gregoire de Nysse*) still remains a remarkably comprehensive summary of Gregory's teaching on the subject. Gregory did not invent the doctrine, but he gave a very personal turn to it, as Fr. Leys says in his Introduction. The Genesis text (1:26), the Wisdom literature of the Jews, St. Paul's use of the concept in the New Testament, Plato, Philo, and Plotinus, the Stoics, Ireneus,

Clement, Origen, Athanasius, and Basil—all these had prepared the way for Gregory.

The personal turn given by Gregory is precisely in making the *Plerōma* or fullness of humanity in space-time *diastēma* the visible manifestation of the being of God in His creation, so much so that the only difference between the original and the image consists in the fact that the original is on one side of the *diastēma* and the image on the other. This means, of course, that the original is self-originate, while the image is derived or created; the original does not have to move from *archē* to *telos* but is always what it wills and wills what it is; the image has to move from beginning to end and achieve perfection through diastematic extension in time-space.

As Hans Urs von Balthasar put it in his poetic study, humanity is the *Presence* of God in *diastēma*. God's *ousia* which is *uncreated* cannot be seen in Man because the latter is *creation*. But in terms of qualities, the created entity called humanity is a reflection of the uncreated entity we call God. And between the original and the image there is no need to posit some kind of intermediary as Plotinus and others would do.

This is Gregory's *tour de force*. Latin thinkers like Tertullian and Hippolytus as well as a North African like Origen, were too much enslaved by the deep-rooted assumptions of Greek thought which posited the logos as a kind of intermediary between the Supreme Being and the Created Order. Gregory, who is wrongly accused of slavish hellenism, did away with this kind of mediation by removing all gradation of bigger and smaller in the Godhead, and by showing the Logos to be fully, integrally and essentially God. There is only one *ousia* of God and the Son is not an instrument of God but fully of the *ousia* of God. Gregory has no use for Plato's world of ideas or Plotinus' *nous* as intermediary. If Gregory speaks of Plato's *kosmos noētos* it is not as intermediary, but as the angelic world of intelligible beings which are part of the created order, and not intermediaries between Creator and creation.

The climax of the *tour de force* comes in positing humanity as the true intermediary between the Creator and the Creation. This is a purely Christian *demarche*, not yet fully understood by Christians in the East or in the West. It is Christ recapitulating in himself the fullness (*pleroma*) of humanity who becomes the right mediator between God and World. This is humanity's true role, and in Christ, humanity is to fulfill this role.

The goal of human existence is not simply to get the Beatific Vision, which 'satisfies' our spiritual desires by making us possess the good. The good, which is God, is never to be possessed like that, according to Gregory. No vision, mystical or otherwise, will enable us to 'possess' God. This desire to possess the good is only projection of our greed and acquisitiveness, our basic sinfulness. The good is not a 'property' to be 'possessed', but the true good is to be in the right relation.

It is in his *Commentary on the Song of Songs* that Gregory has given us his understanding of this movement towards fulfilment, and there, no vision is final. Any vision of God yielded on the upward journey of man towards God, cannot comprise a beholding (*theōria*) or a clear vision (*katalēpsis enargē*) of the Truth, but only the hearing of the voice of the Bridegroom (*phonē tou pothoumenou*)[3]. This is where such excellent scholars like Cardinal Danielou and W. Völker tend to misinterpret Gregory by trying to assimilate his thought into the categories of Western mystical theology.[4] God is not to be grasped in the mystical vision at all, for every time that the soul is embraced by the presence of the Bridegroom, the soul finds that she cannot hold the presence, but has to move on in pursuit. The soul participates only to grow capable of greater participation[5] and so to move on infinitely, for the Good is infinite and cannot be possessed. The voice of the Bridegroom beckons on, and we keep following, keep running. There is no stopping in the race, Gregory tells us, for we move on from glory to glory as the Apostle also taught us. Everytime the soul thinks that it has caught God in its embrace, it finds that God is outside the embrace, moving on, beckoning in love. The Apostle is right, "the eye has not seen, the ear has not heard, nor has the heart of man

---

*3. *In Canticum Canticorum* Oratio V 2:3 GNO VI: 138/9 ff PG 46: 860 A ff.

4. Western 'mystologians' have spoken of various modes or fashions of the Vision of God—to know God in the creature, to know God in one's own soul and to know God in ecstatic union. Von Balthasar (*Presence et Pensee* p. 67 note) has shown how artificial this distinction is, and how the first two are both seeing God in the creature, one in the object and the other in the subject, while the third is merely a more intense experience of the same vision through the veil of creation. Even in the highest mystical experience. Gregory would insist, man encounters only the *energeia* of God, not his *ousia*.

5. *In Canticum Canticornm* V:2:13. GNO VI: 158/15 ff PG 44-876 A.

conceived" (1 Cor. 2:9). What we see is only the boundary of the greater step that lies beyond[6].

If this is our destiny, namely infinite progress in the good and infinite growth in the knowledge of God, how is that to be reconciled to the other pole of our existence in history—i.e. the pole of the physical creation to which also we are related?

This is probably the point at which Gregory has not given us an answer, and we will have to formulate our own. Gregory tells us, however, that the thing that matters is not a so-called Beatific Vision that ends the pilgrimage, but the love of God that keeps propelling us towards a greater and more intimate knowledge of him. Moses, who is supposed to have seen God and received the tablets of Law from God's hands, hangs around and expresses the desire: "I pray Thee, Show me Thy glory" (Exodus 33:18). Yahweh replies "you cannot see my face". (33:20). It is not the seeing that matters, but the love that propels.

That love is the relationship which makes humanity human. As Gregory of Nyssa puts it in his *Commentary on the Song of Songs:* "Is not the name by which Thy goodness is made known, the relation that my soul has to Thee? And how can I but love Thee, who has so loved me, even when I am so black, that you would lay down your own life for Thy sheep which Thou shepherdest?"[7]. And the speaker here is not the individual soul, but the lost sheep—the whole human race, the *plērōma* of humanity in time and space.

How does humanity keep that precarious balance between the two poles—the relationship of God characterized by love, by abiding in Christ and by the indwelling of the Spirit on the one hand, and on the other, the historical relationship to the created order, as the "sensible presence" of the loving God in its midst? The second pole we can truly maintain only when we are truly the image, and in the measure in which we are the image, of the Beloved. Without goodness, and the perfection of goodness, there is no image.

Our relationship to the created order can then not be rectified by mere science and technology. Science and technology themselves can come into their own only when humanity truly becomes the image of God by being perfect in every virtue. Only in the

6. *In Cant. Cant. Oratio* VIII. GNO VI:247/10 ff. PG 44:941 B-C.
7. *In Cant. Cant. Oratio* II:1:7. GNO VI:61/16 ff. PG 44:801 A-B.

measure in which we have *metousia* or participation in the divine perfection can we become the "sensible presence" of God in the Universe, and only then does our science and technology become rectified and our relation to the rest of creation become healthy and health-giving to the creation.

Here we have the profundity of Gregory's thought coming to light, in that dialectic which demands more than 'knowledge' of truth or 'possession' of the good in order to achieve the end of man. Only when humanity *becomes* what it is, i.e. when it manifests its true nature as the Image of God, as the visible manifestation of the divine presence in the creation, does it become fully human. This is not to confuse the *ousia* of God and the *energeia* of God, or the original and the image; created humanity truly manifests itself as the image of God, the visible presence of God in creation.

Hence Gregory's definition of Christianity: "*Christianismos esti tēs Theias Phuseōs mimēsis*"[8] (Christianity is the imitation of God's nature). Gregory does not use the word *Theōsis* in this connection, but the idea is clearly there. Being assimilated to the nature of God does not mean going out of the "measure" of the nature of humanity, for so to go out would be to cease to exist. But the very nature of humanity is to be like God, for that is what it means to be created in the image of God. The more humanity becomes like God, the more it becomes itself. Divinisation is humanisation. *Theōsis is anthrōpēsis.*

But this is the grace of God. Gregory breaks out in lyrical poetry when he sings the praises of the grace of God in his *oration VI* on the Beatitudes:

> "Who is worthy of rendering sufficient thanks for such grace! With what voice, with what understanding, with what movement of delight, can we sing the praises of such super-abundant grace! Humanity virtually goes out of its own nature—becoming immortal from mortal, imperishable from perishing, eternal from ephemeral, becoming wholly God from what is human! For he who is made worthy of being the Son of God, has in every way in himself, the worthiness of the Father, and becomes the inheritor of all the parental goods! O what

8. *De Professione Christiana:* PG 46:244 C.

> magnificent gift from the immeasurably rich Lord! O the breadth of his generosity! O what a generous hand! O the measure of the gifts from that overflowing treasury!"[9]

The God-Man relationship alone can humanize man and make him what he is expected to be in creation—the visible presence of God Himself. The only way to heal humanity as well as the civilization created by its Science and Technology, is for humanity to seek to rectify the other pole of his existence, so that he is not master of the universe in his own right, but only as the image of the Invisible God.

But how does humanity seek its own proper nature, which is the Image of God? We can only indicate the few stages which Gregory recognizes as necessary for the recovery of the Image.

Liberation from Evil

The first stage of course is liberation from the evil with which our nature is now mixed. Image of God means the sum of all good. Evil is thus the opposite of the Image of God. Any mixture of evil thus distorts (but not destroys) the Image.

The beginning of such liberation is Baptism—at this point Gregory's soteriology is quite traditional and conservative. To be baptized is to enter that new reality that the Incarnate Son of God has brought into human nature by his assuming the *plērōma* of human nature into himself. It is in Jesus Christ that human nature becomes totally unmixed with evil, and it is in Him that the new humanity has begun with the Death and Resurrection of Christ. It is from Him alone that humanity as a whole receives the grace and the power to become assimilated to the Image of God, for He alone is the true Image of God. What true human nature is, we see only in Him. The perfect relationship of Image to Original is materialized and manifested in the human life of the Incarnate Christ. The Union of the Divine and the Human in Christ is the historic manifestation of human nature in its "saved" from, for this union, this *henōsis* of humanity with God is an indissoluble one, unlike the relationship of the first Adam to God. In the first Adam the link Original-Image, so essential to human nature, was precariously fragile. Here the Original, the

9. *De Beatitudinibus:* Oratio VII. PG 44:1280 C.

God-head is so inseparably united to body, mind and soul of humanity that there emerges a new humanity which is not only redeemed, but also insured against all loss.

But Baptism is only a beginning. As Gregory would say in his sermon on the Baptism of Christ:

> "The man that was before Baptism was wanton, greedy, acquisitive of what belongs to others, revilers, false, sycophants, and all that sort of thing. Let him now become disciplined (*kosmios*), sober, content with what he has and distributing from it to those that are in poverty, loving the truth, respectful of others, pleasant in his dealings—in short following every praiseworthy course of conduct".[10]

As Gregory goes on to say, darkness is dispelled only by light and the way to drive out evil is to practice goodness. We have to contemplate the characteristics of our Father as true children, and should fashion ourselves according to the image.

Baptism, says Gregory, is initiation to a two-fold practice—the practice of virtue and the practice of worship. The practice of virtue has itself two aspects. On the one hand we have to develop the characteristics of our Father, so that the image in us slowly comes to light. At the same time we have to engage in armed combat against the wiles of the devil. Evil is a force that gives battle, unlike ordinary darkness which flees at the approach of light. The fight against evil and injustice is an essential part of the living out of baptismal grace.

And so has the life of worship two aspects, as Gregory makes clear in the same sermon on the Baptism of Christ. The first is the initiation into the transcendent community—to be knit together with the rest of creation in the world and above the world. The second is our vicarious or priestly task of praising and interceding with God on behalf of the whole creation.

Now where we need to go a step beyond what Gregory explicitly says is to see the whole meaning of baptism in a more corporate sense. It is not enough that the individual gives up greed and acquisitiveness, flattery and deceit, in order to become generous and liberal, disciplined and honest. Such a transforma-

10. *In Diem Luminum* (*In Baptismum Christi*) PG 46:596 C. GNO IX:236. NPNF J:523 B.

tion has to take place on a corporate basis. The Church, and where possible, society as a whole, has to be reconstructed. A new economics, a new politics, a new social education—this is what baptismal grace has to result in. There has been our greatest failure, especially in so-called Christian societies. We failed to make use of the opportunity to create a new economics based not on greed and acquisitiveness, but on disciplined hard work in order to serve the people. We have failed to create political institutions which do not require flattery and dishonesty, and which attracts the most unselfish and the most gifted to positions of leadership. We have failed to create a system of education that removes selfish greed, eliminates love of comfort and creates the social motivation to serve the people.

In other words, baptismal grace was not worked out in our corporate life, even when Christianity was the accepted basis of some of our societies. The Image of God which should have gradually come to light in the whole of humanity, was neglected by the Christian community. A false understanding of mission as the assimilation of other people into a culture dominated by a special race of Christians distorted the proclamation of the Gospel in the world.

Neither has the community's worship achieved the two things. We had partial success at one time in the first aspect—in our identification with and participation in the transcendent community in heaven and earth. But even that participation was marred by racialism, by ethnic and cultural pride, and by over-bearing domination of other Christians by some.

Cosmic Man becomes the Divine Presence in Creation thus in three ways.

(a) the transformation of persons by separation from evil and cleaving to the good;

(b) the transformation of society itself in such a way that the social reality of humanity reflects the *plērōma* of human adherence to the good;

(c) the transfiguration that goes on through the work of the Spirit in Person, Society and *plērōma*, (the whole of humanity) the outcome of which will manifest itself only on That Day.

APPENDIX I

Let Us Not Deviate

"The faith of the Christians, proclaimed to all nations in accordance with the command of the Lord by his disciples, does not have its origin in man, nor is man its cause (Gal 1:1). Rather it has come from and through Jesus Christ our Lord, who, being Logos and Life, Light and Truth, God and Wisdom and all else that He is by nature, for this cause above all was made in the likeness of men, participating in our nature, like us in all things except sin—like us in that he assumed the whole man, body with soul, in order that salvation may be in both these elements of ours. He thus 'was seen on earth and had concourse with human beings' (Baruch 3:38), so that men may no longer need to rely on their own notions for their view of the self-existent, making up doctrines out of conjectures and guess work. Rather, convinced that truly God was manifested in the flesh, let us believe this alone to be the 'mystery of Godliness', the faith as handed over to us by the Logos-God Himself, in direct face to face conversation with the Apostles. Let us then receive the teaching concerning the Transcendent Nature, given to us as through a mirror and as though in a paradox (*ainigmatos*), through the most ancient Scriptures concretized as the Law, the Prophets, and Proverbial wisdom, as a witness to the Truth revealed to us reverently (*eusebōs*) apprehending the sense of the words, in accord with the faith (*hōs...........pistei*) set forth by the Lord of All, let us guard that faith in literal and untampered purity, regarding the slightest deviation from the words traditionally handed down as the highest blasphemy and impiety. We then believe exactly as the Lord set forth the faith to his disciples, saying 'Proceed and teach all the nations, baptising them in the name of the Father and the Son and the Holy Spirit'. This is the word of mystery, by which, through the birth from above, our nature is transfigured from the corruptible to the incorruptible, renewed from the 'old man' in accordance with the image of Him who created in the beginning the likeness to the Godhead. We make then neither subtraction nor alteration or addition, in this faith handed over to the Apostles by God Himself, knowing very well that anyone who dares to pervert the voice of God by evil sophistique is of his father the devil, who departing from the Truth speaks words from himself becoming thus the father of falsehood. For whatever deviates from the truth is falsehood in everyway and not the truth. (*Refutation Confessionis Eunomii*. 1-3. GNO II, pp. 312-313. PG 45. 465-468. NPNF V, 101).

APPENDIX II

Letter of Origen to Gregory Thaumaturgus (Excerpts):

Rejoice in God, O my most excellent master and venerable son, Gregorius.

As you know, the natural gifts of the mind to accomplish some task, can, when sufficiently disciplined, lead to the attainment of that which is denied, in a certain measure, to attain this final end, if I may so name it, on behalf of which the discipline was exercised. Your good natural gifts can make you an excellent Roman lawyer or a Hellenic Philosopher of one of the famous schools. But I would wish that all the power of your natural gifts is directed and put to work towards the goal of the Christian faith.

As for the means to be used, I would suggest that you take from Hellenic philosophy all that you serve as secondary education or academic preparation for Christianity,* and also all those things which are useful in even Geometry and Astronomy for the interpretation of the Holy Scriptures. And thus, as the philosophers say that education in geometry, music, grammar, rhetoric and astronomy can be auxiliary to philosophy, we will also say the same about philosophy, that it is auxiliary to Christianity."

(Then follows an exegesis of the Exodus story of God asking the Israelites to borrow the gold ornaments of the Egyptians before leaving Egypt. It was this gold which then was used to adorn the holy of holies in the tabernacle of Jahweh. Then Origen continues.)

"Why should I make this seemingly inopportune digression? Because I want to make clear for what purpose the objects received from Egypt were useful for the sons of Israel—objects which the Egyptians did not know how rightly to use, but which in the hands of the Hebrews, by the wisdom of God, became useful for the true worship of God.

"The Divine Scripture knows, however, that the descent of the sons of Israel from their own land into Egypt had evil consequences.

---

*Origen uses the term, *ta hoionei eis Christianismon dunamena, genesthai egkuklia mathēmata hē propaideumata,* S.C. 148, pp. 186-188.

It indicates how life among the Egyptians, by which I mean living in the midst of the intellectual disciplines of this world, had evil consequences for those who had been nourished by the law of God and healed by the Israelitic worship of God. Ader the Idoumean, for example, as long as he lived in the land of Israel, never having tasted the food of the Egyptians, was never tempted to make idols. (But when he went down to Egypt, he made idols).

"And for my part, I have learned by experience and can tell you that rare indeed are the people who learned the use of the things received from Egypt and rightly utilized them for the worship of God. But many are the brethren of Ader the Idoumean. These are they who make use of the Hellenic ideas they have picked up to fabricate heresies—so to speak manufacture idols of gold in Bethel (which means House of God). To me this means that they deposited their own creations inside the Scriptures—which is where the Word of God dwells and is allegorically called Bethel. The Scripture says that they had established another idol in Dan. Dan signifies the boundary region, close to the pagan frontiers as is clear from the writings of Joshua, son of Nun. And there, near to pagan territory, are the idols fabricated by the brothers of Ader, of whom we have spoken.

"But you, my lord and my son, be attentive chiefly to the reading of the Scriptures, yes, be attentive, for we need a lot of attentiveness in reading the divine Scripture, in order that we may not say anything improper about it or have inappropriate ideas about it. And being attentive to the reading of the divine books with a view to believing and to be well pleasing to God, knock, while reading, at the doors which are closed; and he will open to you—the gate-keeper of whom Jesus spoke: 'to him the gate-keeper opens' (Jn 10:3). In being attentive to the divine reading, search for the meaning of the divine writings, often hidden to many minds, with straight-forwardness of spirit and with unwavering faith in God. Do not be satisfied merely with knocking and searching; for it is absolutely necessary to pray in order to understand divine matters. . . ."

(Excerpts from letter of Origen to Gregory Thaumaturgus. Greek text ed. H. Crouzel, *Sources Chretiennes*, Vol. 148, Paris 1969. Engl. Tr. present author.)

A SELECT BIBLIOGRAPHY

I. TEXTS

1. Aubineau, M. — *Traité de la Virginité* Sources Chretiennes, 119 Paris 1966
2. Danielou, J. — *Vita Moysis.* Sources Chretiennes, I Paris 1968
3. Forbes, G.H. — *Gregorii Nysseni Basilii fratris quae supersunt omnia tom I.* Fasc. 1-2. Burntisland. 1855-1861
4. Jaeger, W. — et. alii. *Gregorii Nysseni opera,* Leiden, E.J. Brill (For details see pages 21-24)
5. Maraval, P. — *Vita Macrinae* S.C. Paris, 1971
6. Migne, J.P. — *Patrologia Graeca*, Vols. 44, 45, 46.
7. Oehler, F. — *S. Gregorii episc. Nysseni opera* t:I. Halle, 1865
8. Srawley, J.H. — *The Catechetical Oration of St. Gregory of Nyssa,* Cambridge, 1903

II. ENGLISH TRANSLATION OF GREGORY'S WORKS

9. Callahan, Virginia Woods: *Saint Gregory af Nyssa—Ascetical Works,* Catholic University of America Press, Washington D.C. 1967 (Fathers of the Church Vol. 58)
   1) *On Virginity.*
   2) *On what it means to call oneself a Christian (Quid nomen professiove Christianorum Sibi velit)*
   3) *On Perfection*
   4) *On the Christian Mode of Life (de Instituto Christiano)*
   5) *The Life of St. Macrina*
   6) *On the Soul and the Resurrection.*
10. Graef, Hilda C. *The Beatitudes and The Lord's Prayer* in Ancient Christian Writers, 18 (1954)
11. Lowther Clarke, W.K. *St. Gregory of Nyssa, The Life of St. Macrina,* SPCK, London, 1916
12. Moore, W. and Wilson, H.A.W. *Select Writings and Letters of Gregory, Bishop of Nyssa* in *A Select Library of Nicene and Post-Nicene Fathers of the Christian Church,* Second Series, Vol. V. Wm. B. Eerdmans Publishing Company, Grand Rapids, Michigan, 1893.
   Contains the following works:
   1) Contra Eunomium – 12 books
   2) Answer to Eunomius' Second Book
   3) On the Holy Spirit
   4) On the Holy Trinity
   5) On Not Three Gods

6) On the Faith
7) On Virginity
8) On Infants' Early Deaths
9) On Pilgrimages (Epistle I)
10) On the Making of Man
11) On the Soul and the Resurrection
12) The Great Catechism
13) Homily On Meletius
14) On the Baptism of Christ
15) Letters

13. Musurillo, H. *From Glory to Glory, Texts from Gregory of Nyssa* (mostly excerpts from the *Commentary on the Song of Songs*), John Murray, London, 1962.
14. Richardson, C. 1) *An Answer to Ablabius—On Not Three Gods*
2) *An Address on Religious Instruction* in Library of Christian Classics Vol. III., *Christology of the Later Fathers,* ed. E.R. Hardy, London, 1954.
15. Srawley, J.H. *The Catechetical Oration of St. Gregory of Nyssa*—SPCK London, 1917
16. Stein, J.A. *Encomium of St. Gregory, Bishop of Nyssa, on his Brother St. Basil,* Washington, 1928

IV. BOOKS ON ONE OR MORE OF THE ASPECTS OF GREGORY'S THOUGHT

1. ENGLISH

Anderson, A.W. *A comparative study of the concept in St. Gregory of Nyssa's contemplation on the life of Moses and Shri Aurobindo's Commentary in the Veda.* Diss. Columbia Univ. 1960 (pp 163)

Balas, D.L. *Metousia Theou. Man's participation in God's perfections according to St. Gregory of Nyssa.* Stud. Anselm. 55 Roma, Harder 1966 (pp 185)

Cherniss, Harold, *The Platonism of Gregory of Nyssa.* Berkeley, 1930 (pp 92) (Univ. of Calif. Publ. in Class. Philol. XI, 1)

Dunstone A.E. *The Atonement in Gregory of Nyssa. London 1964* (pp. 32) (The Tyndale Press)

Goggin, Sr. Thomas-Aquinas, *The times of St. Gregory of Nyssa as reflected in his Letters and the 'Contra Eunomium'.* Patristic Studies 79, Washington 1947. (Reviews: Botte in *Rech. Theol Anc. et Med.* (1955) p. 346. Greenslade in *Class. Review,* Oxford (1949). Bardy in *Rev. d'hist. ecol.* (1948) p. 553 etc.

Hennesy, J.E. *The background, sources, and meaning of divine infinity in St. Gregory of Nyssa.* Diss. Fordham Univ. 1963 (314 pp)

Hoey, George W.P. *The use of the Optative Mood in the works of Gregory of Nyssa.* Patristic Studies 26. Washington 1930.

Jaeger, Werner. *Two Rediscovered Works of Ancient Christian Literature: Gregory of Nyssa and Macarius of Egypt.* Leiden, E.J. Brill, 1954 (pp. 301).

McGRATH, C. *Gregory of Nyssa's doctrine on knowledge of God.* Diss. Fordham Univ. 1964 (pp. 181).

MESSENGER, Ernest Charles, *Evolution and Theology, The problem of Man's Origin.* London, Burns, Oates & Washbourne, 1931 (XXX+313 pp) Cf. pp. 23-26; 121-144 (Reviews: Dom Butler in Downside Rev. 1932. R.W. Meagher in *Ushaw Magazine*, 1932. W.J. McGarry in *Thought* X, 1935).

MESSENGER, E.C. *Theology and Evolution (A Sequel to Evolution and Theology)*, by various writers, ed. by Messenger, Ph.D. (Louvain). London, Sands & Co. 1950 (337 pp) Cf. pp. 12-21, 87-101.

PFISTER, J.E. *St. Gregory of Nyssa, Biblical Exegete, An Historico-theol. Study of Pentateuchal Exegesis.* Diss. Woodstock (Md. 1964).

PHILIPS, A.J. *The Eschatology of St. Gregory of Nyssa.* Diss. OXFORD, 1963.

SHEA, Joannes, M. *The Church According to St. Gregory of Nyssa's Homilies on the Canticle of Canticles.* Dissertatio ad Lauream. Pontifical Univ. Gregoriana. Baltimore, 1968.

SHELDON-WILLIAMS, J.P. *The Greek Christian Platonist Tradition from the Cappadocians to Maximus and Eriugena.* The Cambridge History of Later Greek and Early Mediaeval Philosophy. Ed. by A.H. Armstrong. Part VI. Cambridge. The University Press, 1967 (711 pp.) Cf. pp. 447 456.

TRINICK, John, *St. Gregory of Nyssa and the Rise of Christian Mysticism*, Shorne (Kent) 1950 (35 pp).

WEISWURM, Alcuin A. *The Nature of Human Knowledge according to St. Gregory of Nyssa.* Diss. Washington 1952. Cath Univ. of Am. Philos. Studies Vol. 136. Wash. C U A Press 1953 (XX, +250 pp) (Reviewed by Leys in *Rev. Phil. Louv.* 1955. p. 427 f. and Orbe in *Gregorianum* 1956. pp 130 f.).

2. GERMAN

AUFHAUSER, Johann Baptist, *Die Heilslehre des Hl. Gregor von Nossa.* Muenchen 1910

BAYER, Johannes, *Die Trostreden des Gregorius von Nyssa in ihrem Verhaeltnis zur antiken Rhetorik.* Marburg 1892. (Theol. Hab. Schr. u. zgl. Lic. Diss).

BAYER, Johannes, *Gregors von Nyssa Gottesbegriff.* Diss. Giessen 1935 (38 pp).

BUSE, A. *Der Heilige Gregor von Nyssa, Leipzig,* 1848.

CAVALLIN, C. *Studien zu den Briefen des heiligen Basilius.* Lund, 1944.

DANIELOU, Jean, *Vom Geheimnis der Geschichte.* Stuttgart, Schwabenverlag O.J. (1956) (404 pp) (Ubers. v. 'Essai sur le mystere de l'histoire', Paris, 1953).

DANIELOU, Jean, *Moses bei Gregor von Nyssa. Vorbild und Gestalt. Moses in Schrift und Ueberlieferung.* Duesseldorf, Patmos, 1963 (330 pp)

DIEKAMP, Franz, *Die Gotteslehre des Hl. Gregor von Nyssa. Ein Beitrag zur Dogmengeschichte der patristischen Zeit.* Muenster i.W. 1896.

DIEKAMP, Franz, *Ueber den Ursprung des Trinitaetsbekenntnisses.* Rede beim Antritt des Rektorats der Westf. Wilh. Univ. gehalten am 15. Okt. 1910. Muenster i. W.O.J.

GOMES, De Castro, M. *Die Trinitaetslehre des heiligen Gregors von Nyssa.* Freiburg i, B. Herder 1938 (XVI, +126 pp) (Reviews: B. Capella in *Rech. Thèol. Anc. et Mèd.* 1940. 179 ff. P. Galtier in Greg. 1940, 293 ff. K. Rahner in *Zeitschr. f. kath. Theol.* 1939. 247. Lebreton in *Rech. Sc. Rel.* 1940. 248-250.

GRONAU, Karl, *Poseidonios aud die juedisch-christ-liche Genesisexegese.* Leipzig 1914 (cf. pp. 112-141).

HILT, Franz, *Des heiligen Gregors von Nyssa Lehre vom Menschen systematisch dargestellt.* Diss. Tuebingen, Koeln 1890.

HOLL, Karl, *Amphilochius von Ikonium in seinem Verhaeltnis zu den grosen Kappadoziern,* Tuebingen und Leipzig 1904 (cf. pp. 196-235).

JAEGER, Werner, *Gregor von Nyssa's Lehre vom Heiligen Geist.* Aus dem Nachlas hrsg. von H. Doerrie: Leiden. E.J. Brill. 1966 (X+153 pp).

KNACKSTEDT, H. *Die Theologie der Jungfraulichkeit beim.* Hl. Gregor von Nyssa, Diss, Rom 1940.

KONSTANTINOU, Evangelos G. *Die Tugendlehre Gregors von Nyssa im Verhaeltnis zu der antik-philosophischen und judisch-christlichen Tradition.* In: *Das Oestliche Christentum.* Neue Folge, Heft 17 Wuerzburg (Augustinus-Verl.) 1968 (167 pp).

KRAMPF, A. *Der Urzustand des Menschen nach der Lehre des Hl. Gregor von Nyssa.* Wuerzburg 1889. Trier 1925 (107 pp).

LENZ, J. *Jesus Christus nach der Lehre des heiligen Gregor von Nyssa. Eine dogmengeschichtliche Studie.* Trier 1925 (123 pp).

MAIER, Johannes, *Die Eucharistielehre der drei grossen Kappadozier.* Diss. Breslau 1915.

MAY, Gerhard, *Gregor von Nyssa und der Abschluss des trinitarischen Dogmas.* Diss. (masch.-schr.) Wien 1964.

MERKI, Hubert, *Homoiosis Theoe, Von der platonischen Angleichung an Gott zur Aehnlichkeit bei Gregor von Nyssa.* Paradosis. Von By Beitraege zur Gesch. der altchristl. Lit u. Theol. 7. Fribourg (S), Paulus, 1952 (XX 188 pp).

MEYER, Wilhelm, *Die Gotteslehre des Gregor von Nyssa. Eine philosophische Studie aus der Zeit der Patristik.* Diss. Jena, Halle 1894.

MUHLENBERG, Ekkehard, *Die Unendlichkeit Gottes bei Gregor von Nyssa. Gregors Kritik am Gottesbegriff der klassischen* Metaphysik. Forschungen zur Kirchen-u. Dogmengesch. 16. Gottingen, Vandenhoeck & Ruprecht 1966 (215 pp).

PREGER, F. *Die Grundlagen der Ethik bei Gregor von Nyssa.* Diss. Leipzig, Wurzburg 1897 (55 pp).

RECHEIS, Athanas, *Engel, Tod und Seelenreise. Das Wirken der Geister beim Heimgang des Menschen in der Lehre der Alexandrinischen und Kappadokischen Vaeter.* Temi e Testi 4. Rom 1958 (222 pp).

REICHE, A. *Die kunstlerischen Elemente in der Welt-und Lebensanschauung des Gregor von Nyssa,* Diss. Jena 1897 (60 pp).

RIEDEL, Wilhelm, *Die Auslegung des Hohen Liedes in der judischen Gemeinde und der griechischen Kirche.* Leipzig 1898 (Cf. pp. 66-74.

RITTER, Heinrich, *Geschichte der Philosophie. 6. Teil.* Hamburg 1841 1841 (Cf. pp. 80-150).

RUPP, Julius, *Gregors des Bischofs von Nyssa Leben und Meinungen.* Leipzig 1834.

SCHLINDER, A. *Die Begrundung der Trinitatslehre in der eunomianischen Kontroverse,* Zurich 1964.

STAATS, Reinhart, *Gregor von Nyssa und die Messalianer, Die Frage der Prioritat zweier altkirchlichscher Schriften.* Patristische Texts und Studien 8. Berlin, Walter de Gruyter & Co, 1968 (144 pp).

STAATS, Reinhart, *Der Traktat des Gregor von Nyssa 'De instituto christiano'. Beweis seiner Abhangigkeit vom Groben Brief des Symeon von Mesopotamien.* Diss. Gottingen 1964 (126 pp) Referat uber diese Diss. in Theol. Lit, zeitung 90 Leipzig 1963 Sp. 791-792, und in Stud. theol. 17, Lund 1963, (p. 120-128).

STIGLER, J.N. *Die Psychologie des Hl. Gregor von Nyssa systematisch dargestellt.* Regensburg 1857.

STIGLMAYR, J. *Makarius der Grobe uud Gregor von Nyssa.* In Theol. u. Glaube, 2, Paderborn 1910 (571 pp).

UNTERSTEIN, K. *Die naturliche Gotteserkenntnis nach der Lehre der Kappadozischen Kirchenveter, Basilius, Gregor von Nazianz und Gregor von Nyssa.* Straubling 1902/1903 (2. Programm des k. hum. Gymn. Straubling).

VOELKER, Walther, *Gregor von Nyssa als Mystiker.* Wiesbaden 1955 (Franz Steiner-Verl.) (XVI+295 pp) Reviewed by Jaeger in *Dt. Lit. Ztg.* Berlin 1957 p. 498-499.

VOLLERT, W. *Die Lehre Gregore von Nyssa vom Guten und Bosen und der schliesslichen Ueberwindung des Bosen.* Leipzig 1897.

WEISS, M. *Die grossen Kappadozier, Basilius, Gregor von Nazianz und Gregor von Nyssa als Exegeten.* Braunsberg 1872.

WEISS, Karl, *Die Erziehungslchre der drei Kappadozier. ·Ein Beitrag zur patristischen Paedagogik.* Strasburger theol. Stud 5, Freiburg I.B. 1903.

ZEMP, Paul, *Die Grundlagen heisgeschichtlichen Denkens bei Gregor von Nyssa, Muenchener theol. Stud. 38.* Munchen (Max Hueber—Verl) 1970 (XX+254 pp).

ZOUBOS, A.N. *Eine Frage des Anaxagoras in der Anthropologie des Gregors von Nyssa.* Athen 1956.

3. FRENCH

AUBINEAU, Michel, *Biographie de Gregoire de Nysse avant 371.* Cap. I de la Introduction a son edition de 'Traitè de la Virginité. Paris 1966. pp. 29-82.

———*Le 'De virginitate' de Grêgoire de Nysse.* Positions des thèses du troisiéme cycle soutenues devant la Faculté en 1963. Publ. de la Fac. des Lettres et Sciences hum. de Paris. Série "Recherches" 25. Paris 1964.

BALTHASAR, Hans Urs v. *Présence et Pensée. Essai sur la philosophie religieuse de Grégoire de Nysse.* Paris (Beauchesne) 1942 (152 pp).

BOUEDRON, P. *Doc'rines psychologiques de saint Grégoire de Nysse.* Nantes 1861 (787 pp).

DAMS, Th. *La controverse eunoméenne.* Diss. (typed) 1951 (Cf Abramowski in *Reallex. f. Ant. u. Chrt.* 'Eunomios'. Et. Daniélou "Eunome l'arien et l'exègèse neéoplatonicienne du Cratyle" en Rev. Sc. Grecques 69 (1958) pp. 412-432.).

DANDELOT, P. *La doctrine du 'diastēma' chez Grégoire de Nysse.* Mémoire licence philos. et lettres. Louvain 1960 (typed).

DANIELOU, Jean, *Essai sur le mystère de l'histoire.* Paris, Seuil, 1953 (346 pp.)

———*L'être et le temps chez Grégoire de Nysse.* Leiden. E.g. Brill 1970 (232 pp.)

DANIELOU, Jean, *Le IVe siècle, Grégoire de Nysse et son milieu.* Notes prises par les éléves. Cours-Confèrence Fac. de Théol. de l'Institut Cath. de Paris. 1965 (188+8 pp.) (Mimeographed).

DANIELOU, Jean, *Platonisme et Théologie Mystique. Essai sur la doctrine spirituelle de saint Grégoire de Nysse.* Coll. 'Théologie' n. 2, Paris, Aubier, 1944 (339 pp.) Nouvelle Edition 1953 (362 pp.) (Reviews: J. Lebreton in *Rech. sc. Rel.* (1946) pp. 249-251. E.V. Ivanka in *Zeitschr. f. kath. Theol.* (1949) pp. 231-233. Ghellink in *Nouv. Rev. Théol.* (1947) p. 330. W. Volker in *Theol. Lit. Ztg.* (1949) pp. 143-148). E. Bréhier in *Rev. Philos.* Paris (1945) pp. 174-176 etc.)

EYNDE, C. Van den, *La version syriaque du Commentaire de Grégoire de Nysse sur le Cantique des cantiques, ses origines, ses témoins, son influence.* Louvain Bureau du Muséon, 1939. vol. XI (135 pp) (Reviewed by Connolly, R.H. in *Jl. Theol. Stud.* 41 (1940) pp. 84-86.

GAITH, J. *La conception de la liberté chez Grégoire de Nysse.* Etudes de Philosophie mèdiévales 43. Paris, Vrin, 1953 (214 pp.)
(Reviews: Leys in Rev. phil. Louv. (1953) pp. 638-639. Hornus in *Et. theol. et rel.* (Montpellier 1956) pp. 79-80. Gilbert in *Nouv. Rev. Théol.* (1956) p. 91. Brunner in *Rev. de Théol. et de Phil.* (Lausanne 1953) p. 286. Caturelli in *Arkhé* (Cordoba, Arg. 1954) pp. 150-153).

GROSS, J. *La divinisation du chrétien d'après les Pères grecs.* Paris 1938. Cf. pp. 219-238.

LEBOEUF, C. *L'unité de la nature divine chez saint Grégoire de Nysse.* Lucioni 1937.

LEYS, Roger, *L'image de Dieu chez saint Grégoire de Nysse. Esquisse d'une doctrine.* Bruxelles L'Edition universelle. Paris Desclée de Brouwer, 1951 (146 pp.) Museum Lessianum sect. théol. 49, Louvain. Nauwelaerts, 1951.
Reviews: Rahner in *Zeitschr. f. kath. Theol.* 1952, 117. Chadwick in *Jl Theol. Stud.* 1952, 270).

LOSSKY, Vladimir, *Vision de Dieu. La tradition patristique* grecque *jusqu'a Grégoire Palamas.* Neuchatel, Delachaux et Niestlé, 1962 (140 pp.)

MACCAULEY, George, *La méthode théologique de Grégoire de Nysse.* Thése pour le Doctorat des Sciences Religeuses. Univ. de Strasbourg. Fac. Thèol. Cath. 1965.

MERIDIER, L. *L'influence de la seconde sophistique sur l'oeuvre de Grégoire de Nysse. Thése Rennes 1906.*

PUECH, Aimè, *Histoire de la littérature grecque crétienne.* Tome III: Le IVe Siécle. Paris 1930. Cf. pp. 396-436.

RIVIERE, Jean, *Le dogme de la rédemption.* Paris 1905, 2. ed.+1933, 3. ed. pp. 151-159; 384-387.

SLOMKOWSKI, Antoine, *L'état primitif de l'homme dans la tradition de l'Eglise avant saint Augustine.* Strasbourg 1928.

TILLEMONT, Sebastien Lenain de, *Mémoires pour servir á l'histoire ecclésiastique.*

4. OTHER LANGUAGES

AKULAS, A. *E peri athanasias tės psuchê. dóxa tou Plátonos en sunkrisei prós tên Gregoriou tou Nusses.* Diss. Jena, Athen 1888 (385 pp).

BERGADES, Ioannes K. *E peri tou súmpantos kaî tês psuchês tou anthrópou didáskalia Gregoriou tou Nússes.* Diss. Leipzig, Thessaloniki 1876 (92 pp).

DE BOER, Sibbele, *De anthropologie van Gregorius van Nyssa.* Academisch procfschrift. Assen 1968 (488 pp).

DIEKAMP, Franz, *Doctrina patrum de incarnatione verbi.* Munster i.W. 1907.

GONZALES, Severino, *La formula 'mia ousia treis hypostáseis' en San Gregorio de Nisa.* Diss Roma 1939. Analecta Gregoriana vol XXI. Series Fac, Theol. Roma Univ. Greg. 1939 (162 pp) (Reviews: Amand de Mandieta in *Rev. Béméd* 1940. pp. 165-166. A. Lieske in *Scholastik* 1941. pp. 264-265. The Camelot in Les *Sc. Phil. et Theol.* 1941-42. pp 448 ff).

GRONAU, Karl, *De Basilio, Gregorio Nazianzeno Nyssenoque Platonis imitatoribus.* Diss. Goettingen 1908.

HERRMANN, Guilielmus, *Oregorii Nysseni sententiae de salute adipiscenda.* Halle 1875 (Diss. theol. habil.)

HEYNS, Stephanus Petrus, *Disputatio historico-theologica de Gregorio Nysseno.* Diss. theol. Leiden. Lugduni Batavorum, Apud J.C. Cyfveer 1835.

JANINI CUESTA, J. *La antropologia y la medicina pastoral de San Gregorio de Nisa.* Madrid, Cons. Sup. Inv. Cientificas, 1945. Reedicion 1946 (202 pp).

KLEINHEIDT, L.S. *S. Gregorii episcopi Nysseni doctrina de angelis.* Freiburg i.B. 1860 (52 pp.)

KOPERSKI, V. *Doctrina Sancti Gregorii Nysseni de processione Filii Dei.* Diss. Rom 1936.

MOELLER, E.W. *Gregorii Nysseni doctrinam de natura hominis et illustravit et cum Origeniana comparavit.* Diss. Halle 1854 (126 pp.)

MOUTSOULAS, Elias, *Gregorios of Nússes os ermeneutés tès Agais Graphés.* Melanges B. Bellas, Athen 1968.

———*E sárkosis tou Lógou kai e théosis toú anthrópou katá tén didaskalian Gregoriou toú Nússes.* Ciss. Athen 1965.

MPONE, K. *Lógos Katechetikós Gregorîou Nusses* Ekdosis phoitetikou theologikou *Sundésmou.* Athenai 1961 (polugr.)

VINCENZI, Alois, Sancti Gregorii Nysseni et Origenis de aeternitate poenarum in vita *futura cum dogmate Catholico omnimoda concordia.* ROMA 1864. Pars I.

5. PERIODICAL ARTICLES

ALEXANDER, Monique, *Le 'De Mortuis' de Grégoire de Nysse.* In: Studia Patristica 10 (Texte und Untersuchungen dur Geschichte der altchristlichen Literat urs. Bd. 107) Berlin 1970. pp 35-43.

ALTANER, Berthold, *Augustinus, Gregor von Nazianz und Gregor von Nyssa. Quellenkritische Untersuchungen.* In: Rév. Bènèd. 61 (1951) pp. 54-62.

AMAND DE MENDIETA, Dom David, *Les ètats de texts des homèlies pseudobasiliennes sur la création de l'homme.* In: Rév. Bénèd. 59 (1949). pp. 3-54.

ANONYMOUS, *Homèlie IX sur le Cantique.* In: Rèv. de la Table Ronde (May 1945). pp. 49-59.

APARICIO DIAZ, F. *El conocimiento mistico en San Gregorio de Nissa bajo las influencias platonicas.* In: Oriente 3, Madrid (1953) pp. 259-266.

ARMSTRONG, Arthur Hilary, *The nature of man in St. Gregory of Nyssa.* In: The Eastern Churches Quarterly 8 (London 1948) pp. 2-9.

———*Platonic elements in St. Gregory of Nyssa's doctrine of man* In Dominic. Stud. 1 (1948) pp. 113-126.

———*The theory of the non-existence of matter in Plotinus* and the Cappadocians. In: Stud. Patr. V (3rd Intern. Patr. Conf. Papers of Oxford 1959), Berlin 1962. pp. 427-429.

AUBINEAU, Michel, *Project de lexique du vocabulaire de* Gregoiré de Nysse. In: Rev. des Et. Grec. LXXV. Paris 1962 pp. XIX-XXI.

BAERT, E. *Le thème de la vision de Dieu chez St. Justin, Clément d'Alexandrie et St. Grégoire de Nysse.* In: Freib. Zeitschr. f. Philos. u. Theol. 12 (1965) 439-497.

BAKER, Alfred, *The Great Letter of Pseudo-Macarius and Gregory of Nyssa.* In: Stud. Monastica (Montserrat) VI (1964) 381-387.

———*Pseudo-Macarius and the Gospel of Thomas.* In: Vigiliae Christianae 18. Amsterdam 1964 215-225.

———*Pseudo-Macarius and Gregorius of Nyssa.* In: Vig. Christ. 20. Amsterdam 1966. 227-234.

BALAS, David. L: *Christian Transformation of Greek Philosophy. Illustrated by Gregory of Nyssa's Use of the Notion of Participation.* Proceedings of the Americ. Cath. Philos. Assoc. 40 (1966) 152-157.

———*Participation in the specific nature according to Gregory of Nyssa. Aristotelian Logic or Platonic Ontology?* In: Actes du Quatriéme Congrés International de Philosophie Médiévale (1967) Univ. de Montréal. Montréal, Paris, 1969. pp. 1079-1085.

BALTHASAR, Hans Urs v., *La Philosophie religieuse de St. Grégoire de Nysse.* Rech. Sc. Rel. 29 (1939) pp 513-569 (pp. 101-139 en "Présence et Pensée").

BAPHEIDES, Nikolaos, *Ai ontologikai kai gnosiologikai theoriai tou en agiois patros emon Gregoriou tou Nusses.* Gregorios Palamas 38. 1955. pp. 128-137. 221-226.

BARDY, G. *Saint Gregoire de Nysse, Ep. ad Philippum.* In: Rech. de Sc. Rel. 11. Paris 1921. pp. 220-222.

BEBIS, G. *Gregory of Nyssa's 'De vita Moysis'. A Philosophical and Theological Analysis.* In: Greek Orthodox Theol. Review 12 (1967). pp. 369-393.

BERGADA, Maria Mercedes, *La concepcion de la libertada en el 'De hominis opificio' de Gregorio de Nyssa.* In: Stromata 24 (1968) pp. 243-263.

———*Elementos aristotelicos en el 'De hominis opificio' de Gregorio de Nyssa.* In: Caud. de Filosofia (Buenos-Aires) 1970. In preparation.

———*El puesto del hombre en el cosmos en el 'De hominis opificio' de Gregorio de Nyssa.* In Eidos 2 (Cordoba, Arg.) 1970. pp. 47-68.

BOUCHET, J.R., *A propos d'une image christologique de Gregoire de Nysse.* In: Rev. Thom. 67 (1967) pp. 584-588.

———*La vision de l'èonomie du salut selon saint Gregoire de Nysse. In Rev.* des sc. phil. et thèol. 52 (Paris 1968) pp. 613-644.

———*Le vocabulaire de l'union et du rapport des natures chez saint Gregoire de Nysse.* In: Rev. Thom. 68. Paris 1968) pp. 553-582.

BOUVY, Edmond, *Sainte Macrine.* In: Revue Augustinienne 1 (1902) pp. 265-288.

BRIGHTMAN, F. *Six notes.* (A text-critical problem in Gregory's 'Antirhet'.) In: Jl. of Theol. Stud. 29 (1928) pp. 158 ff. See Note IV.

BURN, Ronald, *Adversaria in Gregorium Nyssenum.* In: Jl. of Theol. Stud. 25 (1924) pp. 172-177.

———*Adversaria in Gregorium Nyssenum et Macarium Magneten.* In: Jl. of Theol. Stud. 23 (1922) pp. 64-67.

CALLAHAN, J.F. *Greek Philosophy and the Cappadocian Cosmology* In: Dumbarton Oaks Papers XII (Harvard, 1958) pp. 29-57.

———*Gregory of Nyssa and the psychological view of time.* In: Atti del XII Congs. Intern. di Filosofia 12-18 Sept. 1958 (Venezia 1958) vol. XI. pp. 59-66 (Storia della filosofia antica emed. Firemze, Sansoni, 1960.

———*The Serpent and 'e rachia' in Gregory of Nyssa.* In: Traditio 24. (New York 1968) pp. 17-41.

CANEVET, Mariette, *Nature du mal et èconomie du salut chez Gregoire de Nysse.* In: Rech. de Sc. Rel. 56 (Paris 1968) pp. 87-95.

———*Le vocabulaire de la connaissance chez Grègoire de Nysse.* In: Epektasis. Mèlanges Patristiques offertes au Cardinal Jean Danièlou, ed. J. Fontaine et C. Kannengiesser, 1972.

CARLINI, A. *Appunti sul reste del 'De mortuis' di Gregorio di Nissa contenuto nel Vaticano gr. 2066.* In: Annali della Scuola Normae Superiore di Pisa 36 (1967) pp. 83-92.

CATAUDELLA, Mariette, *Le 'De instituto christiano' est-il de Gregoire de Nysse? Problemes de critique interne.* Rev. des Et. Grecqu. 82 (1969) pp. 404-423.

CATAUDELLA, Q. *Griogio Pasquali e le epistoledi Gregorio Nisseno.* In: Atene e Roma, Nuova Serie IV (Messina e Firenze 1959) pp. 193-197.

CAVALLETTI, Sofia, *Grigorio da Nissa e la Haggadah.* In: Sefarad 21 (Madrid 1961) pp. 257-266.

CAVARNOS, John P. *Gregory of Nyssa on the Nature of the Soul.* In: The Greek Orth. Theol. Rev. lx (Brookline, Mass. 1955) pp. 133-141.

———*St. Gregory of Nyssa on the Origin and Destiny of the Soul.* In: Belmont (Mass) Institute for Byzantine and Modern Greek Studies, 1956 (11 pp.).

———*The Psychology of Gregory of Nyssa.* Diss. Harvard 1947. Thesis abstract in: Harvard Studies in Class. Philol. 58-59 (Cambr, Mass. 1948) pp. 219-221.

CHERNISS, M.F. *The platonism of Gregory of Nyssa.* In: Univ. of Calif. Publ. in Class. Phil. XI, 1 (Berkeley 1830) pp. 1-92.

CIGUELA, J.M. *Conceptos metafisicos de Gregorio de Nissa.* In: Revista de la Fac. de Hum. de la Univ. Nac. del Nordeste 5 (Resistencia, Arg. 1963) pp. 73-97.

CONGAR, Yves-M.J. *Le thème de Dieu-Créateur et les explications de l'hexaémeron dans la tradition chrétienne.* In: L'homme devant Dieu. Mèlanges offertes au P.H. de Lubac. Coll. 'Thèologie' n. 56. Paris, Aubier 1963. pp. 189-222

CORSINI, Eugenio, *Nouvelles perspective sur le problèmes des sources de l'hexaémeron de Grégoire de Nysse.* In: Studia Patristica I (Second Intern. Conf. on Patr. Stud. Papers, Oxford 1955)=Texte u. Untersuchungen zur Gesch. der altchristi. Lit. 63) Berlin, Akademieverlag 1957. pp. 94-103.

COURCELLE, Pierre, *Grégoire de Nyssa, lecteur de Porphyre.* In; Rev. des Et. Grecq. 80 (1967) 402-406.

———*Témoins nouveaux de la 'région de dissemblance'.* (Platon, Polit. 273 d) In; Bibl. de l'Ecole des Chartes 118 (Paris 1960) pp 20-36.

CROUZEL, H. *Grégoire de Nysse est-il fondateur de la théologie mystique? Une controverse récente.* In: Rev. d'Asc. et de Myst. 33 (Toulouse 1957) pp. 189-202.

DANIELOU, Jean, L 'Adversus Arium et Sabellium' *de Grégoire de Nyssa et l'origénisme cappadocien.* In: Rech. de Sc. Rel. 54 (Paris 1966) pp. 61-66.

DANIELOU, Jean, '*Akolouthia chez Grègoire de Nysse.* In: Rev. Sc. Rel. 27 (Paris 1953) pp. 219-249 (Republished in Jean Danièlou, "L'Etre et le temps chez Grègoire de Nyssa", Leiden, E.J. Brill, 1970).

———*L'apokatastase chez saint Grègoire de Nysse.* In: Rech. Sc. Rel. 30 (Paris 1940) pp. 328-347.

———*L'apotre selon saint Grègoire de Nyssa.* In: Cahiers de Neuilly 7 (Paris 1960) pp. 21-27.

———*Saint Bernard et les Pères grecs.* In: Saint Bernard Thèologien, Actes du Congrès de Dijon 15-17 Sept. (1953). Analecta Sacri ordinis Cisterciensis 9 (1953) fasc. 3-4 pp. 46-5 5.

———*Chrismation prèbaptismale et divinitè de l'Esprit chez Grègoire de Nysse.* In: Rech. Sc. Rel. 56 (1968) pp. 177-198.

———*La chronologie des oeuvres de Grègoire de Nysse.* In: Stud. Patr. VII (Fourth Intern. Conf. on Patr. Stud. Oxf. 1963)—Texte und Untersuchungen 92 (Berlin, Akademieverl. 1966) pp. 159-169.

———*La chronologie des sermons de Grègoire de Nysse.* In: Rev. Sc. Rel. 29 29 (Strasburg et Paris 1955) pp. 346-372.

———*La colombe et la tènèbre dans la mystique byzantine ancienne.* In: Eranos-Jahrbuch 23 (1954) pp. 398-418.

———*Comble du mal et eschatologie chez Grègoire de Nysse.* In: Festgabe Josef Lorte 2 (Baden-Baden 1957) pp. 27-45 (Republished in Danielou, "L'Etre et le temps". Leiden 1970.

———'*Conspiratio*' *chez Grègoire de Nysse.* In; L'homme devant Dieu. Mèlanges offertes au P.H. Lubac. (Paris Aubier, 1964) vol. I pp. 295-308 (Republ. in Danielou, "L'Etre et le temps", Leiden 1970).

———*Dieu chez Grègoire de Nysse.* In: Rèsurrection 15 (Paris 1960) pp. 68-79

———*Un dossier nèoplatonicien sur la mètempsychose chez Grègoire de Nysse.* In: Melanges G. Florowsky.

———*L'ètat du Christ dans la mort d'après Grègoire de Nysse.* In: Histor. Jahrbuch der Goerres-Gesselschaft Muenchen 77 (1958) pp. 63-72.

———*Eunome l'Arien et l'exègèse nèoplatonicienne du 'Cratyle'.* In: Rev. Et. Grecq. 69 (1956) pp. 412-432.

———*L'èvèque d'apres une lettre de Grègoire de Nysse.* In: Euntes docete 20 (1967) pp. 85-97.

———*La Fète des Tabernacles dans l'exègèse patristique.* In: Stud. Patr. 1 (2nd Inter, Conf. on Patr. Stud. Oxford 1955)—Texte und Udters. 63 (Berlin 1957) pp. 262-279.

———*Saint Grègoire de Nysse dans l'histoire du monachisme.* In: Thèologie de la vie monastique. Etudes sur la tradition patristique. (Paris, Aubier, 1961) pp. 131-141.

———*Grègoire de Nysse á travers les lettres de saint Basile et saint Grégoire de Nazianze.* In: Vigil. Christ. XIX (Amsterdam 1965) pp. 31-41.

———*Grégoire de Nysse et le messalianisme.* In: Rech. Sc. Rel. 48 (1960) pp. 119-134.

———*Grégoire de Nysse et le nèoplatonisme de l'Ecole d'Athènes.* In: Rev. des Et. Grecq. 80 (paris 1967) pp. 395-401.

———*Grègoire de Nysse et Plotin.* In: Association G. Budè. Actes du Congrès de Tours et Poitiers, 1953 (Paris 1954) pp. 259-262.

———*Le marriage de Grègoire de Nysse et la chronologie de sa vie.* In: Mèmorial G. Bardy. Rev. Et. Aug. 2 (Paris 1956) pp. 71-78.

———*Le milieu historique. In:* Introduction a son èdition de la 'Vie de Moise', Paris 1955. pp, II-X.

———*Moise, example et figure chez Grègoire de Nysse.* In: Cahiers Sioniens 8 (Paris 1954) pp. 385-400 (Deutsche Ubersetzung: 'Moses bei Gregor von Nyssa. Vorbils und Gestalt. Moses in Schrift und Uberlieferung', Dusseldorf, Patmosverl. 1963, 330 S.)

———*Le mysière du culte dans les sermons de saint Grègoire de Nysse.* In: "Vom christlichen Mysterium", Gesammelte Arbeiten zum Gedachtnis von Odo Casel. Dusseldorf, Patmosverl. 1951. pp. 76-93.

———*Notes sur trois textes eschatologiques de saint Grèoire de Nysse.* In: Rech. Sc. Rel. 30 (1940) pp. 348-356 (Review by H. Ch. Puech in Rev. de l'Hist. des Relig. 123 (1941) (pp. 217-219).

Danielou, *La notion de confins (methorios) chez Grègoire dt Nysse.* In. Rech. Sc. Rel. 49 (1961) pp. 161-187 (Republished in Daniélou, "L'Btre et le temps", Leiden 1970).

———*Philon et Grégoire de Nysse.* In: Philon d'Alexandrie. (Colloques nationaux du centre nat. de la recherche scientifique, Lyons 1966) (Paris 1967. pp. 339-345.

———*Le problème du changement chez Grégoire de Nysse.* In: Arch. de Philos. Rech. et doc. 29 (Paris, Beauchesne, 1966) pp. 323-347 (Republished in Daniélou, "L'Etre et le temps", Leiden 1970).

———*La résurrection des corps chez Grégoire de Nysse.* In: Vigil. Christ. 7 (Amsterdam 1953) pp. 154-170.

———*Le symbole de la caverne chez Grégoire de Nysse.* In: Mullus, Festschrift Theodor Klauser,=Jahrbuch f. Antike u. Christentum. Ergaenzungsband 1, 1964 (Munster, Aschendorff, 1965) pp. 43-51 (Republished in "L'Etre et le temps", Leiden 1970).

———*Le symbolisme cosmique de la croix chez Grégoire de Nysse.* In: La Maison Dieu 75 (Paris 1963) pp. 23-36.

———*La tènèbre lumineuse. Textes de saint Grègoire de Nysse sur la connaissance de Dieu* In: Chercher Dieu (Paris 1943) pp. 100-111.

———*Le traitè 'Sur les enfanis morts prèmaturèment' de Grègoire de Nysse.* In: Vig. Christ. 20 (Amsterdam 1966) pp. 159-182.

———*Les tuniques de peau chez Grègoire de Nysse.* In: Glaube, Geist, Geschichte. Festschrift fur Ernst Benz zum 60. Geburtstage am 17. Nov. 1967, hrsg. Gerhard Muller und Winfried Zelle. Leiden, E.J. Brill, 1967. pp. 355-367 (Republished in Danielou "L'Etre et le temps", Leiden 1970).

———*La typologie de la semaine au IVe siècle.* In: Rech. Sc. Rel. 35 (Paris 1956) pp 382-411.

DARROUZES, J. *Un receuil èpistolaire byzantin, le manuscript de Patmos 706.* In: Revue des Etudes Byzantines 14 (Paris 1956) pp. 87-121.

DIEKAMP, Franz, *Literaturgeschichtlic hes zur Eunomianischen Kontroverse. In:* Byzantinische Zeitschrift 18 (1909) pp. 1-13.

———*Die Wahl von Gregor von Nyssa zum Metropolitan von Sebaste im Jahre 380.* In: Theolog. Quartalschrift 90 (Tubingen 1908) pp. 384-401.

DOELGER, Franz Joseph, *Das Anhangekreuzchen der hl. Makrina und ihr Ring mit der Kreuzpartikel. Ein Beitrag zur religiosen Volkskunde des 4. Jahrhunderts nach der Vita Macrinae des Gregor von Nyssa.* In: Antike u. Christentum 3, (Munster i.W. 1932) pp. 81-116.

DORRIES, H. *Griechentum und Christentum bei Gregor von Nyssa.* Zu H. Langerbecks Edition des Hohelied-Kommentars in der Leidener Gregor-Ausgabe. In: Theolog. Lit. Ztg. 88 (1963) pp. 569-582.

DRASEKE, Johannes. *Zu Gregorios von Nyssa.* In: Zeitschr fur Kirchengeschichte 28 (Stuttgart 1907) pp. 387-400.

———*Grègorios von Nyssa in den Anfuhrungen des Johannes Erigena.* In: Theol. Studien u. Kritiken 82 (Gotha 1909) pp. 530-576.

———*Textverbesserungen zur 'Oratio catechtica'.* (Review of Mèridier's Edition) In: Theol. Lit. Ztg. 33 (Leipzig 1908) pp. 531-534.

———*Uber den Verfasser der Schrift 'Pros Euagrion monachon peri theotetos'.* In: Jahrbuch fur prot. Theol. 8 (1882) pp. 343-384: 553-568.

DUNSTONE, A. *The meaning of Grace in the writings of Gregory of Nyssa.* In: Scottish Journal of Theol. 15 (Edinburgh 1962) pp. 235-244.

EDSMAN, C.M. *Schopferwille und Geburt. Eine Studie zur altchristlichen Kosmologie.* In: Zeitsch. f.d. neutest. Wissenschaft u.d. Kunde der alteren Kirche 38 (Giessen 1939) pp. 11-44.

ERASTOS, B. *Gregorios o Nusses opos blepei o Jaeger.* In: Aktines 18 (Athens 1955) pp. 213-217.

FLOERI, Fernand, *Le sens de la division des sexes chez Grègoire de Nysse.* In: Rev. Sc. Rel. 27 (1953) pp. 105-111.

FLOROVSKY, G. *The Idea of Creation in Christian Philosophy.* In: The Eastern Churches Quarterly, vol VIII (1949) pp. 25 ff.

GRANDILLAC, M.P. de, *Apropos de Grégoire de Nssse.* In: Dieu Vivant 3 (1945) pp. 123-134.

GEBHARDT, Ernst, *Titel und Zeit der Rede Gregors von Nyssa 'In suam ordinationem'.* In Hermes 89 (Berlin 1961) pp. 503-507.

GIET, S. *Saint Basile a-t-il donné une suite aux homélies de l-Hexaeméron?* In: Rech. Sc. Rel. 33 (Paris 1946) pp. 317-358.

GILLET, R. L'homme divinisateur eosmique dans la pensée de saint Grégoir de Nysse. In: Stud. Patr. VI (Third Intern. Conf. Patr. Stud. Oxford 1959)=Texte u. Untersuchungen 81 (Berlin, Akademieverlag 1962) pp. 62-83.

GONZALES, Severino, *La identidad de operacion en las obras exteriores y la unidad de la naturaleza divina en la teologica trinitaria de San Gregorio di Nisa.* In: Gregorianum 19 (Rom 1938) pp. 280-301.

———*El realismo platonica de San Gregorio de Nissa.* In: Gregorianum 20 (Rom 1939) pp. 189-206.

———*El simbolo de S. Gregorio de Nisa y su posicion entre los simbolos de Cappadocia.* In: Gregorianum 19 (Rom 1938) pp. 130-134.

GORDILLO, Mauricio, *La virginidad transcendente de Maria Madre de Dios en San Gregorio de Nisa y en la antigua tradicion de la Iglesia.* In: Estudios Marianos 21 (Madrid 1960) pp. 117-155 .

GRIBOMONT, J. *Le Panègyrique de la virginité, ovevre de jeunesse de Grègoire de Nysse.* In: Revue d'Ascètique et de Mystique 43 (Toulouse 1967) pp. 249-266.

HAIDACHER, Sebastian, *Rede uber Abraham und Isaak bei Ephraem Syrus und Pseudo-Chrysostomus. Ein Exzerpt aus Gregor von Nyssa.* In: Zeitschr. fur kath. Theol. 1905. pp. 764-766 (See also Zeitschr. f.kath. Theol. 25 (1901) p. 367).

HARL, M. et Canevet. M. *Colloquio su Gregorio di Nisa* Parigi Feb. 1968. In: Rivista di storia e letteratura religiosa 4 (1968) pp. 626-627.

HARL, Marguerte, *A propos d'un passage du 'Contre Euno me' de Grègoire de Nysse: 'aporroia' et les titres du Christ en thèologie trinitaire.* In: Rech. Sc. Rel. 55 (1967) pp. 217-226.

HARL, M. *Les trois quarantaines de la vie de Moise, schéma idéal de la vie du moine-évéque chez les Pères Cappadociens.* In: Rev. Et. Grecq. 80 (1967) pp. 407-412.

HAUSHERR, J. *Ignorance infinie (chez Grégoire de N.)* In: Orìentalia Christ. Periodica 2 (1936) pp. 351-362.

HOLL, Karl, *Uber die Gregor von Nyssa zugeschriebene Schrift, 'Adversus Arium et Sabellium'.* In: Zeitschr. fur Kirchengesch. 25 (Stuttgart 1904) pp. 380-398. Reprinted in: Gesammelte Aufsatze zur Kirchengeschichte II (Tubingen 1928) pp. 298-309.

HORN, Gabriel, *L'amour divin. Note sur le mot 'eros' dans S. Grégoire de Nysse.* In: Rev. d'Asc. et Myst. 6 (Toulouse 1925) pp. 378-389.

———*Le 'miroir',* la *'nuée'. Deux manières de voir Dieu d'après S. Grégoire de Nysse.* In: Rev. d'Asc. et Myst. 8 (Toulouse 1927) pp. 113-131.

ISAYE, G. *L'unité de l'opération dans les écrits trinitaires de saint Grégoire de Nysse.* In: Rech. Sc. Rel. 27 (1937) pp. 422-439.

IVANKA, Endre v. *Die Autorschaft der Homilien 'Eis to poiésomen ánthropon kat' eikóna emetéran kai omoiosin. Poseidonios bei den Kappadokischen Kirchenvatern.* In: Byzantinische Zeitschrift 36 (Leipzig 1936) pp. 46-57.

IVANKA, Endre v. *Vom Platonismus zur Theorie der Mystik (Zur Erkenntnislehre Gregors von Nyssa)* In: Scholastik XI (Freiburg 1936) pp. 263-195.

———*Die Quelle von Ciceros 'De natura deorum II, 45-60. (Poseidonios bei Gregor von Nyssa).* In: Archivum Philologicum 59 (Budapest 1935) pp. 3-12.

———*Die stoische Anthropologie in der lateinischen Literatur*. In: Anzeiger der Oesterreich. Akad. der Wissenschaften 87 (1950) pp. 178-192.

JAEGER, Werner, *Vom Affen und wahren Christen*. In: Varia variorum. Festgabe Karl Reinhardt 14. Feb. 1951 (Munster, Bohlau-Verlag, 1952) pp. 161-168.

———*Die asketisch-mystische Theologie des Gregor von Nyssa*. In: W. Jaeger, "Humanistische Reden und Vortrage". Berlin, de Gruyter, 2. Aufl. 1960. pp. 266-286.

———*Besprechung von: Merki, Hubert, "Omoiesis theo" In: Paradosis 7, Fribourg 1952*. In Gnomon 27 (1955) pp. 573-580.

———*Der neuentdeckte Kommentar zum Johannesevangelium und Dionysios Areopagites*. In: Sitzungsberichts Berlin 1930. Phil. hist. pp. 569-594.

JAININI CUESTA, J. *Dieta y virginidad: Basilio de Ancira y San Gregorio de Nisa*. In: Miscel. Comillas 14 (1950) pp. 187-197.

———*La penitencia medicinal desde la Didascalia Apostolorum a San Gregorio de Nisa*. In: Rev. Esp. Theol. 7 (Madrid 1947) pp. 337-362.

JOLLY, R. *Sur deux thèmes mystioues de Grègoire de Nysse*. In: Byzantion 36 (Bruxelles 1966) pp. 127-143.

KALLISTOS, Archim, *O hágios Gregorios o Nusses*. In: Nèa Sion 40 (Jerusalem 1945) pp. 110-124; 206-227.

KALSBAGH, Adolf, *Vom Mensch-und Lebensideal der christlichen Antike. Gregore von Nyssa 'Vita S. Macrinae'*. In: "Das Bild vom Menschen". Beitrage zur theologischen and philosophischen Anthropologie, Fritz Tillman zum 60. Geburtstag gewidmet von Schulern und Freuden, Dusseldorf, Verlag L. Schwann, 1934 (249 pp). See esp. pp. 36-41.

KANNENGIESSER, Charles, *L'infinite divine chez Grègoire de Nysse*. In: Rech. Sc. Rel. 55 (1967) pp. 55-65 (summarises work of E. Muhlenberg, "Die Unendlichkeit Gottes bei Gregor von Nyssa")

KAPENEKA, I. *Eisègesis eis ton 'Katechetikon Logon ton mègan' Gregoriou Nusses*. In: Nèa Sion 50 (Jerusalem 1955) pp. 80-87; 183-190; 395-400. 51 (1956) pp. 33-42; 129-135.

KEENAN, Mary Emily, *St. Gregory of Nyssa and the medical profession*. In: Bull. Hist. of Medicine 15 (Baltimore 1944) pp. 150-161.

———*The Letters of St. Gregory of Nyssa, In Class.* Weekly 37 (New York 1943-44) pp. 75-77.

———*'De Professione Christiana' and 'De Perfectione'. A study of the Ascetical doctrine of St. Gregory of Nyssa*. In: Dumbarton Oaks Papers V (Cambr. Mass. Harvard U.P. 1950) pp. 167-207.

KEMMER, Alfons, *Gregorius Nyssenus estne inter fontes Joannis Cassiani numerandus?* In: Orientalia Christana Periodica 21 (1955) pp. 451-466.

———*Gregor von Nyssa und Psuedo-Makarius*. Der Messalianismus im Lichte östlicher Herzensmystik. In: Antonius Magnus Eremita. Studia Anselmiana 38 (Roma 1956) pp. 268-282.

———*Messalianismus bei Gregor von Nyssa und Psuedo-Makarius*. In: Rev. Bènèd. 72 (Meredsous 1962) pp. 273-306.

KIRCHMEYER, J. *Le 'De Virginitate' de Grègoire de Nysse*. In: Science et Esprit 21 (1969) pp. 141-148.

KLIJN, A.F. *Some Remarks on the quotations of the Gospel in Gregory of*

Nyssa's 'De Instituto Christiano' and Makarius' Epistula Magna! In: Vigil. Christ. 19 (Amsterdam 1965) pp. 164-168.

KLOPPEL, Makrina, *Makrina die Jungere. Eine altchristliche Frauengestalt.* In: "Frauen im Bannkreis Christi". Gesammelte Aufsatze von Theodor Bogler (Hrsg), Maria Laach, Verlag Ars Liturgica, 1964. (96 pp). Liturgie und Monchtum, Laacher Hefte 35, pp. 80-94.

KOCH, Heinrich, *Das mystische Schauen beim Hl. Gregor von Nyssa.* In: Theol. Quartalschrift 80 (Tubingen 1898) pp. 397-420.

KOTTING, Berhard, *Gregor von Nyssa's Wallfahrtskritik.* In: Stud. Patr. V (Third Intern. Conf. Patr. Stud. Oxford 1959)=Texte u. Untersuchungen 80 (Berlin 1962) pp. 360-367.

KORNARAKES, Ioannes, *E pros Letoion kanonikâ epistolè Gregoriou tou Nusses ex apopseos poimantikes psuchologias.* In: Gregorios a Palamas 42 (Thessalonike 1959) pp. 147-152, 219-231.

KUGENER, M.A. *Une homèlie de Sèvère d'Antioche attribuèe á Grègoire de Nysse et á Hèsychius de Jèrusalem.* In: Rev. de l'Orient Chrètien 3 (Paris 1898) pp. 435-451.

LADNER, Gerhart, *St. Gregory of Nyssa and St. Augustine on the symbolism of the Cross.* In: Late Class and Med. Stud. in Honor of Albert Mathias Friend Jr. (Princeton Univ. Press 1955) 88-95.

———*The philosophical anthropology of St. Gregory of Nyssa.* In: Dumbarton Oaks Papers XII (Cambr. Mass. Harvard U.P. 1958) pp. 59-94.

LAFONTAINE-DOSOGNE, J. *Nouvelles notes cappadociennes.* In: Byzantion 33 (Bruxelles 1963) pp. 121-183.

LANGERBECK, Hermann, *Zur Interpretation Gregors von Nyssa.* In: Theol. Lit. Ztg. 82 (Leipzig 1957) Sp. 81-90.

LA PIANA, Giorgio, *Una omelia inedita di S. Gregorio Nisseno e le omelie 'eis ton euangelismon' attribue a S. Gregorio Thaumaturgo.* In: Rivista storico-critica della Scienze Teologiche 5 (1909) pp. 527-563.

LEANEY, R. *The Lucan Text of the Lord's Prayer.* In: Novum Testamentum I, (Leiden 1956) pp. 103-111.

LEBON, J. *Le sort du consubstantiel niceen. II.* In: Rev. d'Hist. Ecclès. 48 (Louvain 1953) pp. 632-682.

LENDLE, Otto, *Zur Uberlieferung der zweiten Predigt Gregors von Nyssa auf Stephanus.* In: Byz. Zeitschr. 58 (Leipzig 1965) pp. 320-326.

LEVASTI, Arrigo, *La dottrina mistica di S. Gregorio di Nissa.* In: Rivista di ascetica e mistica 12 (1967) pp. 548-552. 13 (1968) pp. 44-61.

LEYS, R. *La thèologie spirituelle de Grègoire de Nysse.* In: Stud. Patr. 2 (2nd Intern Conf. Patr. Stud. Oxford 1855)=Texte u. Unters. 64 (Berlin) 1957 pp. 495-511.

LIESKE, A. *Zur Theologie der Christusmystik Gregors von Nyssa.* In Scholastik 14 (Freiburg 1939) pp. 485-514. Reimpressinn : Munsterischen Beitrage zur Theologie 22, Munster 1943.

———*Die Theologie der Christusmystik Gregors von Nyssa.* In: Zeitschr. f. kath. Theol. 70 (1948) pp. 49-93; 129-168; 315-340 (Reviewed by Danielou in Rech. Sc. Rel. 36 (1949) pp. 617-620.

LOT-BORODINE, M. *L'anthropologie thèocentrique de l'Orient chrètien comme base de son expèrience spiriuelle.* In: Irénikon 16 (1939) pp. 6-21.

———*La doctrine della dèification dans l'Eglise grecque jusqu'au XI siècle.* In: Rev. de l'Hist. des Relig. 53 (1932) pp. 5-43; 525-574. 54 (1933) pp. 8-55.

MAAS, Paul, *Die Echtheit der neuen Briefe Gregors von Nyssa.* In: Berliner Philologische Wochenschrift 1913, pp. 1470-1472.

———*Gregorius von Nyssa und der griechische Ephrem.* In: Byzantinische—neugriechische Jahrbucher 1920. p. 337.

MAROTTA, E. *La base biblica della Vita sanctae Macrinae di Gregorio di Nyssa.* In: Vetera Christianorum 5 (1968) pp. 73-88.

———*L'ironia e altri schemi nel 'Contra Fatum' di S. Gregorio di Nissa.* In: Vetera Christianorum 4 (1967) pp. 85-105.

MARRIOT, G.L. *The 'De Instituto Christiano' attributed to Gregory of Nyssa.* In: Jl. of Theol. Stud. 19 (London 1918) pp. 328-330.

MATEO-SECO, L.F. *La Teologia de la muerte en la 'Oratio catechetica magna' de San Gregorio de Nissa.* In: Scriptura Theologica 1. fasc. 2 (Pamplona 1969) pp. 453-473.

MAY, Gerhard, *Die Datierung der Rede 'In suam ordinationem' des Gregor von Nyssa und die Verhandlungen mit den Pneumatomachen auf dem Konzil von Konstantinopel 381.* In: Vigil Christianae 23 (1969) pp. 38-57.

———*Gregor von Nyssa in der Kirchenpolitik seiner Zeit.* In: Jahrbuch der osterreich. byz. Gesellsch. 15 (Graz, Koln 1966) pp. 105-132.

MCCLEAR, E.V. *The Fall of Man and Original Sin in the Theology of Gregory of Nyssa.* In: Theol. Stud. 9 (Baltimore, Woodstock, 1948) pp. 175-212.

MCCLELLAN, W.H. *Genesis 2:7 and the Evolution of the Human Body. A study in Literal Exegesis.* In: Eccles. Review 72 (Philadelphia 1925) pp. 1-10.

MCDONOUGH, J.A. *The treatise of Gregory of Nyssa's 'Contra fatum'.* A critical text with prolegomena. Rèsumè d'une thèse dans Harvard Studies in Class. Philo. 61 (Cambr. Mass. H.U.P. 1953) pp. 179-180.

MCGARY, W.J. *St. Gregory of Nyssa and Adam's Body.* In: Thought X (1935-36) pp. 81-94.

MERCIER, J. *A l'ècole des Pères Grecs.* In: Cahiers de Neuilly 9 (Paris 1959) pp. 25-35.

MICHAUD, E. *St. Grègoire de Nysse et l'apocatastase.* In: Rev. Intern. Thèol. X (Bern 1902) pp. 27-52 (=Internationale kirchliche Zeitschrift, Bern).

MUCKLE, J. *The Doctrine of St. Gregory of Nyssa on Man as the Image of God.* In: Mediev. Stud. 7 (Toronto 1945) pp. 55-84.

MUHLENBERG, Ekkehard, *Apollinaris von Laodicea.* Gottingen, Vandenhoek & Ruprecht, 1969 (257 pp)=Forschungen zur Kirchen—u. Dogmengesch. No. 23 On Gregory's 'apodeixis' see pages 64-91.

MULLER, Friedrich, *Der zwanzigste Brief des Gregor von Nyssa.* In: Hermes 74 (Berlin 1939) pp. 66-91.

MUSURILLO, Herbert, *History and Symbol. A study of form in early Christian literature.* In: Theol. Stud. 18 (Baltimore 1957) pp. 357-386.

———*The illusion of prosperity in Sophocles and Gregory of Nyssa.* In: American Journal of Philology 82 (Baltimore John Hopkins Press, 1961) pp. 182-187.

———*A Note on Gregory of Nyssa's Commentary on the Song of Solomon. Homily IV.* In: Didascalia. Studies in honor of Anselm M. Alberada, ed. Sestro Prete. (New York, Bernard M. Resenthal Inc. 1961) pp. 321-326.

NICOSIA, Carmela, *Platonisme e pessimismo nel pensiero di Gregorio di Nisa.* In: Miscell. di Studi di Litterat. Crist. antica VI (Catania 1956) pp. 23-35.

ORTH, E. *Curae criticae.* In: Eremita 26 (1958) pp. 201-203.

OTIS, Brooks, *Cappadocian Thought as a coherent System.* In: Dumparton Oaks Papers XII (Cambr. Mass. Harvard U. P. 1958) pp. 95-124.

OWEN, E.C. *St. Gregory of Nyssa. 1) Grammar, 2) Vocabulary and Style.* In: The Jl. of Theol. Stud. 26 (1925) pp. 64-71.

PAPADOPOULOS, I.A. *Gregorios o Nusses ōs rhètor kai didaskalos.* In: Anaplasis 36 (Athens 1923) pp, 10; 22-23; 39; 53; 109-110; 127-129; 140-141.

PARYS, Michel van, *Un colloque sur Grègoire de Nysse.* In: Irenikon 42 (1969) pp. 540-545.

———*Grègoire de Nysse, moine, thèologien et èvèque.* In: Seminarium 2 (Citta del Vaticano 1969) pp. 292-312.

PASQUALI, Giorgio, *Le lettre de Gregorio di Nisa.* Studi Italiani di Filologia Classica 3 (Florenz 1923) pp. 75-136.

PELIKAN, Jaroslav, *The Mortality of God and the Immortality of Man in Gregory of Nyssa.* In: The Scope of Grace. Essays on Nature and Grace. In honor of J. Sittler. Ed. Philip J. Hefner. Philadelphia, Fortress Press 1964) pp. 77-97.

PELLEGRINO, M. *Il platonismo die San Gregorio Nisseno nel dialogo intorno all' anima e alla risurrezione.* In: Riv. Fil. Neoscolast. 30 (Milan 1948) pp. 437-474.

PFISTER, J.E. *A biographical note. The brothers and sisters of St. Gregory of Nyssa.* In Vig. Christ. 18 (Leiden 1964) pp. 108-113.

PHILLIPOU, A.J. *The Doctrine of Evil in St. Gregory of Nyssa.* In: Studia Patr. 9 (Texte u. Unters. 94) (Berlin 1966) pp. 251-256.

PHOKULIDES, Ioannes, *Gregorios o Nusses. In: Panteinos* 28 (1936) pp. 26-27.

QUERE-JAULMES, F. *L'aumone chez Gregoire de Nysse et Grègoire de Nazianze.* In: Stud. Patr. 8 (4th Intern. Conf. Patr. Stud. Oxford 1963)=Texte u. Unters. 93 (Berlin 1966) pp. 449-455.

REBECCHI, *L'antropologia naturale die S. Gregorio Nisseno.* In: Divus Thomas. 46 (Piacenza 1943) pp. 176-195; 309-341.

RENEHAN, R. *Some Greek Textual Problems. De opificio hominis* 27. In: Harvard Studies in Class. Phil. 67 (Cambr. Mass. 1963) pp. 269-283.

RICHARD, M. *L'introduction du mot 'hypostase' dans la thèologie del' incarnation.* In: Mèlanges de Sc. Rel. 2 (Lille, Fac. Cath. 1945) pp. 5-32; 243-270.

RITTER, M. *Gregor von Nyssa. 'In suam ordinationem' Eine Quelle fur die Geschichte des Konzils von Konstantiopel 381.* In: Zeitschr. f. Kirchengesch. 79 (Stuttgart 1968) pp. 308-328.

RIVAS, H.L. *El conocimiento analogico de Dios en los santos Padres* de *Cappadocia.* In: Revista de Teol. 8 (La Plata, Arg. 1959) pp. 68-74.

ROUGIER, Lois, *Le sens des termes 'ousia' 'upostasis' et 'prosopon' dans les controverses trinitaires post-nicèennes.* In: Rev. de. l'histoire des religions 74 (1916) pp. 48-63; 133-189; 175-189.

SADDINGTON, D.B. *The function of education according to Christian writers of the latter part of the fourth century.* In: Acta Classica (Proceedings of the Class. Assoc. of South Africa) VIII (Cape Town 1965) pp. 89-101.

SAJDAK, Ioannes, *De oratione 'eis ton euangelismon' falso Gregorio Nysseno adscripta.* In: Eos 21 (1916) pp. 9-20.

SCHOEMAN, J.B. *Gregors von Nyssa theologische Anthropologie als Bildtheologie.* In: Scholastik 18 (Freiburg 1943) pp. 31-53; 175-200.

SIMON, J. *Ou et quand furent prononcèes les 'orationes in XL Martyres' de s. Grègoire de Nysse?* In: Handes Amsorga, Monatschrift fur armenische Philologie 41 (Wien 1927) pp. 733 ff.

SKOUTERIS, Konstantinos, *'Meteoroporein' kai 'summeteoroporein para to hagio Gregorio Nyssés.* In: Theologia 39 (Athens 1968) pp. 425-439.

SOELL, G. *Die Mariologie des Kappadozier im Lichte der Dogmengeschichte.* In: Theol. Quartalschrift (Stuttgart 1951) pp. 163-188; 288-319; 426-457.

SPIRA, A. *Rhetorik und Theologie in den Grabreden Gregors von Nyssa.* In: Stud. Patr. 9 (Texte und Unters. 94) (Berlin 1966) pp. 106-114.

SPUROPOULOS, G. *Ai peri lèpras kai lepron koinonikai apopseis Gregoriou tou Nusses.* In: Aktines 22 (Athens 1959) pp. 158-162.

SRAWLEY, J.H. *St. Gregory of Nyssa on the sinlessness of Christ.* In Jl. of Theol. Stud. 7 (London 1906) pp. 434-441.

———*The manuscripts and Text of the 'Oratio Catechetica' of St. Gregory of Nyssa.* In: Jl. of Theol. Stud. 3 (London 1902) pp. 421-428.

STAATS, Reinhart, *Die Asketen aus Mesopotamien in der Rede des Gregor von Nyssa 'In suam ordinationem'.* In: Vigil. Christ. 21 (Amsterdam 1967) pp. 165-179.

———*Die Datierung von 'In suam ordinationem des Gregor von Nyssa.* In: Vigil. Christ. 23 (1969) pp. 58-59.

———*Der Traktat Gregors von Nyssa 'De instituto christiano' und der grosse Brief Symeons.* In: Studia Theol. cura Ordinum Theolog. Scandinaviorum 17 (Lund 1963) pp. 120-128.

STEPHANOU, E. *La coexistence initiale du corps et de l'ame d'après saint Grègoire de Nysse et saint Maxime l'Homologète.* In: Echos d'Orient 31 (Paris 1932) pp. 304-315.

STIGLMAYR, J. *Die Schrift des hl. Gregor von Nyssa 'Uber die Jungfraulichkeit".* In: Zeitschr, fur Aszese und Mystik 2 (Innsbruck, Munchen 1927 (pp. 334-359.

STOLPE, S. *Platonism och kristendom. Eros och agape hos Gregorius av Nyssa.* In: Credo 34 (Berlin 1953) pp. 152-157.

SUTCLIFFE, Edmund F. *St. Gregory of Nyssa and Paradise.* In: The American Eccles. Review 84 (Washington 1931) pp. 337-350 (Extract in Messenger, "Theology and Evolution", London 1949) pp. 97-98.

TELFER, W. *The birth of Christian anthropology.* In: Jl. Theol. Stud. n.s. 13 (1962) pp. 347-354.

TRINICK, John, *St. Gregory of Nyssa and the Doctrine of the Image.* In: The East. Church Quart. 9, 4 (1951) pp. 175-184.

VILLECOURT, L. *La grande lettre grecque de Macaire, ses formes textuelles et son milieu littèraire.* In: Revue de l'Orient Chrètien 22 (Paris 1920/21) pp. 29-56.

VISSER, A. *De zesde en de achtste: Gregorius van Nyssa over de zesde psalm.* In: Homiletica en biblica 20 (Den Haag 1963) pp. 126-131.

VOLKER, Walther, *Zur Gotteslehre Gregors von Nyssa.* Vigil. Christ. 9 (1955) pp 103-128.

———*Die Mystik Gregors von Nyssa in ihren geschichtllichen Zusammenhangen.* In: Theol. Zeitschr, 9 (Basel 1953) pp. 338-354.

———*Die Ontologie Gregors von Nyssa in ihren Grundzugen.* In: Festschrift G. Biundo (Grunstadt, Pfalz 1952) pp. 9-16.

———*Virginitat and Apathie als Hohepunkte asketischen Strebens bei Gregor von Nyssa.* In: Festschrift Dymtro Cyzevsky (Berlin 1954) pp. 301-106.

VOLLERT, W. *Hat Gregor von Nyssa die paulinische Eschatologie verandert?* In: Theol. Blatter 14 (Leipsig 1935) pp. 106-112.

WENGER, A. *Grègoire de Nysse et le Psuedo-Macaire.* In: Rev. des Et. Byzant. 13 (Paris 1955) pp 145-150.

WIFSTRAND, A. *The new Edition of Gregory of Nyssa's Commentary on the Song of Solomon.* In: Jl. Theol Stud. n.s. 12 (Oxford 1961) pp. 291-298.

WILAMOWITZ-MOELLENDORF, Ulrich v. *Uber geplante Gregor von Nyssa-Ausgabe.* In: Byz, Zeitschr. 18 (Munchcn 1909) pp. 711 ff.

WILMART, A. *La tradition de l'hypotypose ou traitè sur l'ascèse attribuè a s. Grègoire de Nysse.* In: Rev. de l'Orient Chrètien 21 (1919) pp. 412-421.

WINDEN, J.C.M. van, *Grègoire de Nysse: 'De anima et resurrectionè* PG 46. 17 A. In: Vigil. Christ. 22 (Amsterdam 1968) p. 256.

WOLFSON, Harry Austyn, *The Identification of 'Ex nihilo' with 'Emanation' in Gregory of Nyssa.* In: Harvard Theol. Review 63 (1970) pp. 53-60.

VI. RELATED BOOKS AND ARTICLES

ALBERTZ, Martin, *Untersuchungen uber die Schriften des Eunomius.* Diss. theol. Halle, Wittenberg 1908.

AMAND DE MENDIETA, Dom David, *Fatalisme et libertè dans l'antiquitè Grecque.* Louvain Univ. Bibl. 1945. Cf. pp. 405-439 (ch. X on Gregory of Nyssa).

ARMSTRONG, A. Hilary, *St. Augustine and Christian Platonism.* Villanova Univ. Press 1967 (VIII+66 pp) St. Augustine Lecture Series 1964+66. (Reviewed by Courcelle in Rev. Et. Grecq 82 (1969) pp. 675f).

ARNOU, Renè, *Praxis et Theoria, Etudes de dètail sur le vocabulaire de la pensèe des Ennèades de Plotin.* (Cf. pp. 49-50).

ANWANDER, Friedrich, *Die literarische Bekampfung des Christentums in der Antike.* In: Benedikt. Monatsschrift zur Pflege rel. und geistigen Lebens 6 (1924) pp. 297-320.

AUBINEAU, Michel, *Le thème du bourbier dans la littèrature grecque profane et chrètienne.* In: Rech. Sc. Relig. 47 (Paris 1959) pp. 185-214.

BARTELINK, G.J.M. *Quelques observations sur 'parhesia' dans la littèrature palèo-chrètienne.* In: Graecitas et Latinitas Christianorum Primaeca, Supplementa 3, 1 (Nijmegen 1670) pp. 7-57.

BAUR, L. *Untersuchungen uber die Vergottlichungslehre in der Theologie der griechischen Vaeter.* In: Theol. Quartalschrift 98 (1916) pp. 4467-91.

BEIERWALTER, WERNER, *Plotin uber Ewigkeit und Zeit (Enneads III, 7).* Ubersetzt, eingeleitet und kommentiert von Werner Beirwalter, Frankfurt M.V. Klostermann, 1967.

BENITO Y DURAN, Angel, *San Basilio Magno, punto de partida para el estud'o de la autognoselogia cristiana.* In: Augustinus 6 (Madrid 1961) pp. 315-338.

BLUMENBERG, H. *Kritik und Rezepfion antiker Philosophie in der Patristik: Strukturanalysen zu einer Morphologie der Tradition.* Studium Generale 12, Heidelberg 1959. pp. 485-497.

BOLL, F. *Stndien uber Claudius Ptolemaeus, Ein Beitrag zur griechischen Philosophie und Astrologie, In:* Jahrbucher fur klassische Philologie. Supplementband 21 (1904) pp. 181 ff.

BOURASSA, Francois, *Excellence de la virginitè: Arguments patrisitiques,* In: Sciences Eccles. 5 (Montreal 1953) pp. 29-41.

BRUCK, Eberhard F. *Kirchenvater und soziales Erdrecht, Wanderungen religioser Ideen durch die Rechte der ostlichen und westlichen Welt.* Berlin, Springer, 1956 (XI+286 pp).

BURGHARDT, W.J. *The Image of God in Man according to Cyril of Alexandria.* The Cath. Univ. of America. Stud. in Christian Antiquity 14, Washington 1957.

CALLAHEN, John F. *Augustine and the Greek Philosophers.* Villanova Univ. Press, 1967 (XVI+117 pp).

———*Basil Of Caesarea. A New Source for St. Augustine's Theory of Time.* Harvard Stud. in Class. Phil. 63, 1958. pp. 437-454.

CHERNISS, H. *Aristotle's criticisms of Plato and the Academy.* Vol. I. Baltimore, 1944.

———*Aristotle's critism of pre-socratic philosophy.* Baltimore, 1935.

CLAGHORN, G.S. *Aristotle's criticism of Plato's Timaeus.* The Hague, 1954.

COURCELLE, Pierre, *Traditions nèoplatoniciennes et traditions chrètiennes de la 'region de dissemblance'.* Plato, Politique 273 d. In: Arch. d'Hist. Doctr. et Litt. du M. Age 32 (1957) pp. 5-33.

———*Tradition platonicienne et traditions chrètiennes du corps-prison (Phèdon. 62b, Cratyle, 400 c)* In: Rev. des Et. Latines 43 (Paris 1965) pp. 406-443.

DEHNHARD, H. *Das Problem der Abhangigkeit des Basilius von Plotin, Quellenuntersuchungen zu seinen Schrifteng De Spiritu Sanctu'.* Patrist. Texte u. Studien 3, Berlin, 1964 (100 pp).

DEMAN, Th. *Remarques critique de S. Thomas sur Aristote interprète de Platon,* In: Rech. Sc. Philos. Theol. 30 (1941-42) pp, 133-148.

DEVOS, P. *St. Grègoire de Nazianze et Hcllade de Cèsarèe en Cappadoce.* In: Analecta Bollandiana 79 (1961) pp. 91-101.

———*S. Pierre l_{er}, èvèque de Sebastee dans une lettre de S. Grègoire de Nazianze.* In: Analecta Bollandiana 79 (1961) pp. 346-360.

DES PLACES, Edouard, *Syngeneia. La parentè de l'homme avec Dieu d'Homère á la patristique.* Paris, 1964 (Cf. esp. pp. 195-202).

———*La tradition patristique de Platon.* In: Rev. Et. Grecq. 80 (Paris 1967) 385-394 pp.

DORRIES, Hermann, *Christlicher Humanismus und* monchische Geist-Ethik. In Theol. Lit. Ztg. 79 (1954) pp. 643-656.

———*De spiritu sancto. Der Beitrag des Basilius zum Abschlus des trinitarischen Dogmas.* Abhandlungen der Akad. des Wiss. in Gottingen, Phil. hist. 3. Folge Nr. 39. Gottingen, Vandenhoeck & Ruprecht, 1956 (199 pp) Cf. esp. pp. 10; 85; 133.

DORLODOT, Henry de, *Le Darwinisme au point de vue de l'orthodoxie catholique.* 1er vol. L'origine des espèces. Bruxelles, Paris, Vromant & Co. (1921) (123 pp) Cf. esp. pp. 82-101.

English edition: *Darwinism and Catholio Thought*, London, 1922. Cf. pp. 73-79.

EHRHARDT, A. *The Beginning: a study in the Greek Philosophical Approach to the concept of Creation from Anaximander to St. John.* 1968.

EIDL, Hans, *Augustin und die Patristik* (Geschichte der Philosophie in Einzeldarstellungen, Abt. III, Die Christliche Philosophie Band 10/11) Munchen Verlag Ernest Reinhardt, 1925 (62 pp).

ELLIGER, Walter, *Die Stellung der alten Christen zu den Bildern in den ersten vier Jahrhunderten. Nach den Angaben der zeitgenossischen kirchlichen Schriftsteller.* Leipzig, Dieterich'sche Verlags-buchandlung, 1930 (98 pp).

EUCKEN, Rudolf, *Die Lebensonschauungen der grosen Denker. Eine Entwicklungsgeschichte des Lebens-problems der Menschheit von Plato bis zur Gegenwar.* 17. und 18. Aufl. Berlin und Leipzig, de Gruyter, 1922 (564 pp) Cf. esp. pp. 201-206.

FESTUGIERE, A.J. *Vraisemblance psychologique et forme litteraire chez les anciens.* In: Philologus 102 (1958) pp. 21-42.

GALAN Y GUTIERREZ, Eustaquio, *Ius naturae* (Lecciones de Catèdra). Valladolid, Meseta. 1954 (XVI+652 pp).

GALTIER, Paul *La remission des peches moindres dans l'èglise du troisième au cinquième siècle.* In: Rech. Sc. Rel. 13 (1923) pp. 97-129.

GANSS, George E. *Platonism and Early Christian Thought.* In: Thought XI (New York 1936) pp. 68-85.

GERLITZ, Peter, *Der mystische Bildbegriff (eikon und imago) in der fruhchristlichen Geistgeschichte. Die philosophischen Grundlagen der Enteschatologisierung des altkirchlichen Dogmas.* In: Zeitschr. Rel. u. Geistesgesch. 15 (1963) pp. 244-256.

GHELLINCK, J. de, *Un aspet de l'oppostion entre Hellènisme et Christianisme: L'attitude vis-a-vis de la dialectique dans les dèbats trinitaire.* In: Patr. et Moyen Age III (1948) pp. 245-310.

GIET, S. *De. S. Basile á S. Ambrose. La condamnation du prèt à intèret au IVe siecle.* In: Rech. Sc. Rel. 32 (1944) pp. 95-128.

GILSON, Etienne, *Pourquoi saint Thomas critiquè saint Augustin?* In: Div. Thom. 1 (Frieburg 1926-27).

GOLDSCHMIDT, V. *Le systéme stoicism et l'idèe de temps.* 2nd revd. ed. 1969.

GRAEF, HILDA, *The Light and the Rainbow. A Study in Christian spirituality from its roots in the Old-Testament and its development through the New-Testament and the Fathers to Recent Times.* London, Longmans, Green. Westminster, Md. Newman, 1959 (IX+414 pp) Cf. esp. pp. 147-165.

GRILLMEIER, A. *Patristische Vorbilder fruhscholastischeg Systematik, Zugleich ein Beitrag zur Geschichte des Augustinismus.* In: Stud. Patr. 6 (=Texte u. unters. 81) Berlin 1962. pp. 147-165.

GRONAU, Karl, *Poseidonius, eine Quelle fur Basilius' Hexahemeros.* Braunschweig, 1912.

———*Das Theodizeeproblem in der altchristlichen Auffassung.* Tubingen, J.C. B. Mohr (Paul Siebeck), 1922 (VIII+130 pp).

GUITTON, Jean, *Le temps et l'èternitè chez Plotine et saint Augustine.* Paris, Boivin, 1933.

HACKIN, F. *La Prètendue Passion inèdite de S. Alexandre de Thessalonique.* In: La Nouvelle Clio 6 (Bruxelles 1954) pp. 70-72.

HAMMANN, Adalbert *L'enseignement sur la crèation dans l'Antiquitè chretienne.* In: Rev. Sc. Rel. 42 (1968) pp. 1-23; 97-122.

———*Riches et pauvres dans l'Eglise ancienne.* In: Lettres Chrètiennes, Vol. 6. Paris 1965) (pp. 318).

HANSEN, G. Chr. *Molestiae nuptiarum.* In: Wissensch. Zeitschr. der Univ. Rostock 12 (1963) pp. 215-219.

HARL, Marguerite, *Adam et les deux arbres du Paradis (Gen. II, III) ou l'homme milieu entre deux termes (mèsos-methorios) chez Philon d'Alexandrie (Pour une histoire de la doctrine du libre-arbitre).* In: Rech. Sc. Rel. 50 (1962) pp. 321-388.

———*Le guetteur et la cible: les deux sens de 'skopos' dans la langue religieuse des chrétièns.* In: Rev. Et. Grecq. 74 (Paris 1961) pp. 450-468.

HATCH, Edwin, *The Influence of Greek Ideas on Christianity.* Foreword with new notes and a bibliography by Frederick C. Grant, Harper Torchbooks 18, New York 1957 (359 pp).

HAUSCHILD, Wolf Dieter, *Christliche Trinitaetslehre und griechischer Paideia-Begriff.* Diss. Hamburg. Ev. Theol. Fak. 1967.

HOFFMANN, Manfred, *Der Dialog bei den christlichen Schriftstellern der ersten vier Jahrhunderte.* Texte uud Untersuchungen 96 (Berlin, Akademie, Verlag, 1966) (XX+168 pp) cf. esp. pp. 130-135.

HONIGMANN, Ernst, *Trois mèmoires posthumes d'histoire et de geographie de l'Orient Chrètien.* Hrsg. v. Paul Dėvos. In: Subsidia hagiographica 35 (Bruxelles 1961) pp. 22-35.

IMAMICHI, T. *Die Notizen von der Metamorphose der klassischen Ethik bei den griechischen Kirchenvaetern.* In: Stud. Patr. 5 (Texte u. Unters. 80), Berlin, 1962, pp. 499-507.

IVANKA, Endre von, *Hellenisches und Christlisches im fruhbyzantinischen Geistesleben.* Wien 1948 (119 pp) cf. pp. 28-67 'The Cappadocian Fathers'

———*Plato christianus, Ubernahme und Umgestaltung des Platonismus durch die Vaeter.* Einsiedeln, Johannes-Verlag, 1964 (496 pp) Cf. pp. 151-185.

———*Vom Platonismus zur Theorie der Mystik. (Zur* Erkenntnislehre Gregors von Nyssa). In: Scholastik 11 (Freiburg 1956) pp. 163-195.

———*Zum Problem des christlichen Neuplatonismus: was heist eigentlich christlicher Neuplatonismus?* In: Scholastik 31 (Freburg 1956) pp. 31-40.

———*Vaeterphilosophie und Existenzanalyse.* In: Gloria Dei 9 (1954) pp. 48-51.

JAEGER, Werner, *Aristotle* Second Edition. Oxford, 1948.

———*Early Christanity and Greek Paideia.* Cambr. Mass. The Belknap Press of H.U.P. 1961 (154 pp).

———*Nemesios von Emesa. Quellenforschungen zur Neuplatonismus und seinen Anfaegen bei Poseidonios.* Berlin, Weidmannsche Buchandlung, 1924 (X+148 pp).

———*Paideia. The Ideals of Greek Culture.* Oxford, New York, 1939-44. 3 vols.

———*The Theology of the Early Greek Philosophers.* Oxford 1947

KANNENGIESSER, Charles, *Philon et les Pères sur la double crèation de l'homme.* In: 'Philon d'Alexandriè', Colloques nationaux du CNRS, Lyons 1966. (Paris 1967) pp. 277-296.

KARPP, Heinrich, *Probleme altchristlicher Anthropologie. Biblische Anthropologie und philosophische Psychologie bei den Kirchenvaeern des dritten Jahrhunderts.* Beitrage zur Forderung christlicher Theo. 44,3. (Gutersloh, Bertelsmann, 1950) (256 pp) Cf. esp. pp. 240-243.

KENT, Fr. *Woman in classical and patristic literature.* In: The Tablet 19 (April 1924) London.

KOCH, Hugo, *Die altchristliche Bilderfrage nach den literarischen Quellen.* Forschungen zur Relig. u. Lit. des Alten u. Neuen Testaments, 27 N.F. 10. Gottingen, Vanderhoeck & Ruprecht, 1917 (IV+108 pp).

KONTAGONOS, K. *Philologikè kai kritikè Historia.* Tomos b. Athenai, 1853.

KOTTING, Bernhard, *Peregrinatio religiosa. Wallfahrt und Pilgerwesen in Antike und alter Kirche.* Munster, 1950. cf. pp. 431-424.

LADNER, Gerhart B. *The Idea of Reform. Its impact on Christian Thought and Action in the Age of the Fathers.* Cambr. Mass. Harvard U.P. 1959 (553 pp) cf. pp. 75-78.

———*Die mittelalterliche Reform-Idee und ihr Verhaeltnis zur Idee der Renaissance.* In: Mitteilungen des Inst. fur Oesterreich. Gesh.-forschung 60 (1952) pp. 31-59.

LAFFRANQUE, Marie, *Poseidonios d'Apamée.* Essai de mise au point. Paris 1964 (579 pp).

LEBOURLIER, J. *Apropos de l'etat du Christ dans la mort.* In: Rev. des Sc. Phil. et Théol. 46 (1962) pp. 629-649. 47 (1963) pp. 161-180

LEBRETON, J. *Le dogme de la Trinité dans l'ancienne Eglise Chrétienne.* In: Rech. Sc. Rel. 8 (1918) pp. 1-15

LEIEGANG, Hans, *Die Begriffe der Zeit und Ewig*-keit im spateren Platonismus. Baumkers Beitrage, Ed. 13, Heft 4. Munster, 1913

LEROY, F. *Une homélie mariale de Proclus de Constantinople et le pseudo-Grégaire le Thaumaturge.* In: Byzantion 33 (Bruxelles 1963), Homage a B. Lavagnini. pp. 357-384

LEWY, H. *Sobria Ebrietas. Untersuchungen zur Geschichte der antiken Mystik.* Beihefte zur Zeitschr. fur die neutest. Wissenschaft 9 (Giessen 1929) pp. 132-136

LIETZMANN, Hans, *Apolinaris von Laodicea und seine Schule.* Tebingen, 1904. cf. esp. pp. 83-87

LOGOTHETOS, Konstantinos, *E philosophia ton Patéron kai tou mésou ainos Méros A'.* Athens, 1930. Cf. pp. 236-259 (Reviewed by J. Sebon in *Rev.* d'hist. eccl. 26 (1930) p. 810)

LOSSKY, Vladmir, *Vision de Dieu. La tradition patristique grecoue jusqu'a Grégoire Palamas.* Neuchatel, Delachaux et Niestlé, 1962 (140 pp)=*Schau Gottes.* Zurich EVZ-Verlag, 1964 (Bibl. fur orthodoxe Theol u. Kirche 2)= *The Vision of God.* Faith Press, London

LUBAC, Henri de, *Le Mystère du Surnaturel.* Paris, Aubier, 1965 (300 pp).

LUNEAU, Auguste, *L'histoire du salut chez les Pères de l'Eglise.* La doctrine des ages du monde. Paris, Beauchesne, 1964. Cf. pp. 161-187.

LUTHARDT, Christoph Ernst, *Die Lehre vom freien Willen und seinem Verhaeltnis zur Gnade, in ihrer geschichtlichen Entwicklung dargestellt.* Leipzig, 1863 (XII+470 pp) Cf. pp. 17-20.

MALEVEZ, Léopold, *L'Eglise dans le Christ.* In: Rech. Sc. Rel. 25 (1935) pp. 257-292: 418-440.

MANSION, S. *La critique de la théorie des Idées dans le 'peri ideon' d'Aristote.* In: Rev. philos. Louvain 47 (1949) pp. 169-202.

MARECHAL, Joseph, *Etudes sur la psychologie des mystiques.* Paris, Desclée, 1937. Cf. Tome II, pp. 101-111.

MARKOWSKI, H. *Zum Briefwechsel zwischen Basilios und Libanios.* Berliner Philologische Wochenschrift 1913. pp. 1150-1152.

MASTROGIANNOPOULOS, I.H. *Oi patéres tes Ekklesias kai o anthropos.* Athênai 1966 (396 pp).

MERSCH, Emile, *Le corps mystique du Christ.* 2. Ed. Bruxelles 1936. Cf. pp. 450-463. 3. Ed. Paris 1951.

MOENA, Anselmo de, *A historia teologica da Assuncao* In: Rev. Ecles. Brasil. 10 (1950) pp. 260-272.

MUSURILLO, Herbert, *The problem of ascetical fasting in the Greek Patristical writers.* In: Traditio 12 (1966) pp. 1-64.

NAUTIN, Pierre, *Homélies Pascales III.* Sources Chrétiennes 48, Paris 1957. pp. 84-105.

NORDEN, Eduard, *Die antike Kunstprosa com VI. Jahrhundert v. Chr. bis. in die Zeit der Renaissance.* 2 Bde. Leipzig, Teubner, 1898. Cf. 2. Bd. pp. 558-562.

NORRIS, Richard A. *God and World in Early Christian Thought.* London, A. and C. Black, 1966.

NYGREN, Anders, *Eros und Agape, Gestaltwandlungen der christlichen Liebe.* 2. Teil. (Guetersloh, Bertelsmann, 1937) (605 pp) Cf. pp. 232-251 Translated in French and English.

O'HAGEN, A.P. *Material Recreation in the Apostolic Fathers,* 1968 (=Texte und Untersuchungen 100).

OSBORN, Henry Fairfield, *From the Greeks to Darwin. An outline of the Development of the Evolution idea.* New York, Macmillan & Co., 1894 (259 pp) Cf. p. 71.

OTIS, Brooks, *Nicene Orthodoxy and Fourth Century Mysticism.* Proceedings for the Intern. Congr. of Byzant. Stud. II (Ochrid. 10-16 Sept. 1961) Beograd, 1964. pp. 475-484.

———*Vergil and Clio. A consideration of Vergil's relation to history.* In: Phoenix 20 (1966) pp. 59-75.

PATTERSON, L.G. *The conversion of Diastema in the Patristic view of time.* In: R.A. Norris, Jr. (ed). "Lux in Lumine". Essays to Honor W. Norman Pittenger. New York, Seabury, 1966.

———*God and History in Early Christian Thought. A Study of Themes from Justin Martyr to Gregory the Great.* In Series: Studies in Patr. Thought, London, A. and C. Black, 1967.

PETRE, H. *Ordinata caritas. Un enseignement d'Origéne sur la charité.* In: Rech. Sc. Rel. 42 (1954) pp. 40-57.

PRESTIGE, G.L. *God in Patristic Thought.* London, 2. Ed. 1956.

PUECH, Henri-Charles, *La ténèbre mystique chez les Pseudo-Denys l'Areopagite et dans la tradition patristique.* In: Etudes Carmélitaines mystiques 23 (1938) pp. 33-53.

QUISPEL, G. *The Syrian Thomas and the Syrian Macarius.* In: Vigil. Christ. 18 (Amsterdam 1964) pp. 226-235.

RAHNER, Hugo, *Die Gottesgeburt, Die Lehre der Kirchenvaeter von der Geburt Christi im Herzen der Glaeubigen.* In: Zeitschr. fuer kath. Theol. 59 (Wein 1935) pp. 333-418.

RAVEN, C.E. *Apolinarianism.* Cambridge, 1923. Cf. pp. 262-270.

REFOULE, Francois. *La date de la lettre à Euagre* (PG 46: 1101-1108) In: Rech. Sc. Rel. 69 (Paris 1961) pp. 520-548.

RIVIERE, Jean, *Le marché avec le démon chez les Pères antérieure à saint Augustin.* In: Rev. Ac. Rel. 8 (1928) pp. 257-270.

ROUSSEAU, O. *La rencontre de saint Ephrem et de saint Basile.* In: L'orient Syrien 3 (Paris 1958) pp. 73-91; 86-90.

SCHLINGENSIEPEN, H. *"Die Wunder des Neuen Testaments". Wege und Abwege- ihrer Deutung in der alten Kirche bis zur Mitte des Fuenften Jahrhunderts.* Beitraege zur Foerderung christl. Theol. 2 (Guetersloh, Bertelsmann, 1933) (XII+288 pp).

SCHNEIDER, Arthur, *Die mystisch-ekstatische Gottesschau in griechischen und christlichcn Altertum.* In: Philos. Jahrbuch der Goerres-Ges 31 (1918) pp. 24-42.

SCHWARTZ, Eduard, *Zur Kirchengeschichte des 4. Jahrhunderts.* Gesammelte Schriften, Ed. 4: Zur Geschichte der alten Kirche und ihres Rechts. Berlin, Walter de Gruyter & Co. 1960 1-110.

SINNIGE, T.G. *Matter and Infinity in the Pre-Socratic Schools and Plato*, 1968.

SMORONSKI, Kazimierz, *'Et spiritus Dei ferebatur super aquas'. Inquisitio historico-exegetica in interpretationem textus Gen. 1.2c.* In: Biblica Commentarii editi a Pontifico Instituto Biblico 6 (Roma 1925) pp. 140-156; 275-293; 361-395.

SPANNEUT, M. *Le stoicsme des Pères de l'Eglise.* 1957 In: Series Patr. Sorbon. ed. H.J. Marrow.

STEPHANOU, E. *Le sixième jour de l'Hexaemeron de S. Basile.* In: Echos d'Orient 31 (Paris 1932) pp. 385-398.

STIEGELE, Paul, *Der Agennesiebegriff in der griechischen Theologie des vierten Jahrhunderts. Ein Beitrag zur Geschichte der trinitarischen Terminologie.* Freiburger theol. Studien 12 (Frieburg i.B., Herder, 1913) (XIV+144 pp).

STIGLMAYR, Joseph, *Kirchenvaeter und Klassizismus. Stimmen der Vorzeit ueber humanistische Bildung.* Freiburg i. B., Herder, 1913 (104 pp.) Cf. pp. 5; 16.

STRUCKER, Arnold, *Die Gottesebenbildlichkeit des Menschen in der christlichen Literatur der ersten zwei Jahrhunderte. Ein Beitrag zur Geschichte der Exegese von Gen. 1. 26.* Muenster i. W. 1913 (XI+133 pp).

TATAKIS, Basileios, *E sumbolé tês Kappadokîas stè christianikè skèpse.* Athénai 1960 (285 pp).

THEILER, W. *Diotima neuplatonisch.* In: Archiv fuer Geschichte der Philosophie 50 (1948) pp. 29-47.

———*Forschungen zum neuplatonismus.* In: Quellen und Studien zur Gesch. der Philosophie Bd. X. hrsg. Paul Wilpert. Berlin, de Gruyter, 1966.

———*Untersuchungen zur antiken Literatur.* Berlin, 1970.

TYZIAK, Julius, *Wege oestlicher Theologie. Geistesgeschichtlche Durchblicke.* Bonn, Verlag der Buchgemeinde, 1946.

USENER, Hermann, *Das Weihnachtsfest. Religionsgeschichtliche Untersuchungen.* 2, Aufl. Teil 1. Hrsg. Hans Lietzmann. Bonn, 1911 Cf. pp. 225 ff.

VANDENBUSSCHE, E. *La part de la dialectique dans la théologie d'Eunomius—le technologue*'. In: Rev. d'Hist. Eccles. 40. (Louvain 1944-45) pp. 47-72.

VILLER, Marcel, et RAHNER, Karl, *Aszess und Mystik in der Vaeterzeit.* Freiburg, 1939. Cf. pp. 133-145.

VOLKER, Walther, *Das Volkommenheitsideal des Origenes.* Beitr. z. hist. Theol 7 (Tuebingen 1931).

WALTHER, Georg, *Untersuchungen zur Geschichte der griechischen Vaterunserexegese.* Texte und Untersuchungen 40, 3 (Leipzig 1914) pp. 31-49.

WATSON, G. *The Stoic Theory of Knowledge,* 1966.

WEISS, H F. *Untersuchungen zur Kosmologie des hellenistischen und palaestinlschen Judentums.* Texte und Untersuchungen, 97, 1966.

WELSERSHEIME, L. *Das Kirchenbild der griechischen Vaeterkommentare zum Hohenlied.* In: Zeitschr. fuer kath. Theol 70 (Wien 1948) pp. 423-433.

WILPERT, Paul, *Die philosophische Kritik im Altertum und Mittelalter.* Studium Generale 12 (1959) pp. 471-485.

WILSON, R. McL. *The Early History of the Exegesis of Gen. 1 : 26.* In: Stud. Patr. 1 (1957) pp. 420ff.

WOLFSON, H.A. *The Eternity of the World.* In: Harvard Theol. Review 59 (1966) pp. 352ff,

———*Philo: Foundations of Religious Philosophy in Judaism, Christianity, and Islam.* 2 vols. Cambr. Mass. 1947.

———*The Philosophy of the Church Fathers.* Vol. I. *Faith, Trinity, Incarnation.* Harvard U.P. 1956 (XXVIII+635 pp).

WOOLLCOMBE, K.J. *Le sens de 'type' chez les pères.* In: Vie spiritualle. Suppl. tom. 5 (Paris 1951) pp. 84-100.

VII. REFERENCE BOOKS AND ENCYCLOPAEDIA ARTICLES

ARNOU, R. *Le Platonisme des Pères.* In: Dict. Thèol. Cath. XII, 2 (Paris 1935) col. 2258-2392.

BARDENHEWER, O. *Geschichte der altkirchlichen Literatur.* Bd. III (Freiburg 1911) pp. 188-220 Neue Aufl. in Freiburg, 1924.

———*Gregor von Nyssa.* In: Kirchenlexikon, 2. Aufl. (Freiburg i.B. 1888) Sp. 1167-1177.

BARDY, G. *Macédonius et les Macédoniens.* In: Dict. Theol. Cath. 9 (Paris 1927) col. 1464-1478.

BERNARD, A. et Le Bras, G. *Usure I.* In: Dict. Theol. Cath. 15 II (Oarus 1948) col. 2316-2372.

BOUYER, Louis, *La spiritualité du N.T. et des Pères.* In: Histoire de la spiritualitè chrétienne. vol. I (Paris, Aubier, 1960) Cf. Ch. XIV et XV.

CANEVET, Mariette, *Saint Grégoire de Nysse.* In: Dictionnaire de Spiritualité 6 (Paris, 1967) col. 971-1011.

DANIELOU, Jean, *Gregorio Nisseno, Santo.* In: Encicl. Cattolica 6 (Citta del Vaticano, 1951) col. 1096-1111.

———*Mystique de la ténèbre chez Grégoire de Nysse.* In: Dict. de Spir. II (1953) col. 1872-1885.

GHENYN, Y. van den, *Grégoire de Nysse.* In: Dict. de la Bible, t. III (Paris 1912) pp. 334-335.

GODET, P. *Grégoire de Nysse.* In: Dict. Theol. Cath. 6 (Paris 1920) col. 1847-1852.

GRAF, G. *Geschichte der christlichen arabischen Literatur,* Bd. I. Studi e Testi. Publicazione della Biblioteca Vaticana 118 (Roma 1944). Cf. esp. pp. 332-335.

GRAF, H.C. *Gregorios von Nyssa.* In: Lexikon fuer Theologie und Kirche. 2. Aufl. IV (Freiburg 1960) Sp. 1211-1213.

KOLLIAS, El. *Gregorios o Nusses, Eikonographèia.* In: Threskeutikē kai ethikê enkuklopaideia 4os tomos (Athénai 1964) pp. 771-772.

KONSTANTINIDES, E. *Gregorios o Nusses.* In: Threskeutikê kai ethikê enkuklopaideia 4 (Athénai 1964) pp. 754-771.

KURIAKOU, D. *Encheiridion Patrologias.* Athénai 1898.

LADNER, G.B. *Ernuerung.* In: Reallexikon fuer Antike und Christentum Bd. VI (Stuttgart 1966) pp. 240-275.

LOOFS, F. *Gregor von Nyssa.* In: Realenzyklopaedie fuer Prot. Theol. und Kirche, Bd. VII (3. Aufl. Leipzig 1899) pp. 146-153.

LOTH, B. Michel, A. *Grégoire de Nysse St.* In: Dict. Théol. Cath. 16 (Paris 1959) Sp. 1935-1937.

MEINHOLD, P. *Pneumatomachoi.* In: Realencyclopaedie der klassischen Altertumswissenschaft 21 (Stuttgart 1951) pp. 1066-1101.

PETERS, F.E. *Greek Philosophical Terms.* A historical Lexicon, 1967.

QUASTEN, Johannes, *Patrology Vol. III. The Golden Age of Greek Patristic Literature from the Council of Nicaea to the Council of Chalcedon.* Utrecht, Antwerp, Westminster, Md. 1963. On Gregory of Nyssa: pp. 254-295. On the Cappadocians: pp. 203-295.

SAUGET, J.M. *Gregoiro di Nissa.* In: Bibliotheca Sanctorum 7 (1966) co. 205-210.

UEBERWEG, Freidrich, *Grundris der Geschichte der Philo ophie der patristischen und scholastischen Zeit.* Zehnte, Vollstaendig neu bearbeitete und stark vermehrte. Auflage. Hrsg. Matthias Baumgartner (Berlin, E.S. Mittler und Sohn, 1915) (658 pp. Supple. 266 pp). On Gregory: pp. 129-138 and Supplement pp. 62-63.

VENABLES, E. *Gregorius (15) Nyssenus.* In: A Dict. of Christian Biography, ed. W. Smith and H. Wace, vol. II (London 1880) pp. 761-768.

INDEX